Dampness in Dwellings

This revised and updated book provides a definition of dampness in each of its forms; it details the various potential sources and causes that can result in damage to the building and damage to the health of the occupiers. It is both practical and provides an outline of the possible solutions, looking at aspects of building design and construction that can reduce or avoid the risk of dampness. It also discusses why dampness is a risk to the health of occupiers and so justifies the need to protect health by reducing or removing it.

This book:

- Provides a valuable resource for housing, legal, and medical professionals
- Discusses possible solutions in aspects of building design and construction that can reduce or avoid the risk of dampness and also the safe removal of mould
- Provides an explanation of the legal framework in the UK and potential legal remedies for adverse health effects as a result of dampness

The new edition of *Dampness in Dwellings* includes expanded details on the health implications of dampness in the home, legal updates, and new thinking in the wake of the tragic death of Awaab Ishak. It provides a pivotal resource for active professionals in housing, medical, and legal sectors.

Dr Stephen Battersby, MBE, is an environmental health practitioner, independent consultant, and advisor and was part of the team that developed the Housing Health and Safety Rating System at Warwick University.

Dr Véronique Ezratty, MD, is a medical doctor, health risk assessor in the Group Health, Safety and Prevention Department (DP2SG) of EDF (Electricité de France).

Routledge Focus on Environmental Health
Series Editor
Stephen Battersby
MBE PhD, FCIEH, FRSPH

Regulating the Privately Rented Housing Sector
Evidence into Practice
Edited by Jill Stewart and Russell Moffatt

Dampness in Dwellings, 1st Edition
Causes and Effects
David Ormandy, Véronique Ezratty and Stephen Battersby

Tackling Environmental Health Inequalities in a South African City?
Rediscovering Regulation, Local Government and its Environmental Health
Practitioners
Rob Couch

Leadership Lessons from a Global Health Crisis
From the Pandemic to the Climate Emergency
Jo Nurse

Sewerage and Sewage as an Environmental Health Issue
Stephen Battersby

Private Sector Housing and Health
Evaluating the Effectiveness of Regulation Intended to Protect the Health of
Tenants
Paul Oatt

Dampness in Dwellings, 2nd Edition
Causes, Effects and Remedies
Stephen Battersby and Véronique Ezratty

For more information about this series, please visit: https://www.routledge.com/
Routledge-Focus-on-Environmental-Health/book-series/CONENHE

Dampness in Dwellings
Causes, Effects and Remedies
Second Edition

Stephen Battersby and Véronique Ezratty

Routledge
Taylor & Francis Group
LONDON AND NEW YORK

Chartered Institute of
Environmental Health

Second edition published 2025
by Routledge
4 Park Square, Milton Park, Abingdon, Oxon, OX14 4RN

and by Routledge
605 Third Avenue, New York, NY 10158

Routledge is an imprint of the Taylor & Francis Group, an informa business

© 2025 Stephen Battersby and Véronique Ezratty

First edition published by Routledge 2020

British Library Cataloguing-in-Publication Data
A catalogue record for this book is available from the British Library

ISBN: 978-1-032-95859-0 (hbk)
ISBN: 978-1-032-95858-3 (pbk)
ISBN: 978-1-003-58685-2 (ebk)

DOI: 10.1201/9781003586852

Typeset in Times New Roman
by Apex CoVantage, LLC

This book is the second edition of that published in 2022 and is dedicated to the memory of David Ormandy, who wrote much of the first edition. The proposal to update the work was submitted before his death. For much of his working life he had sought to use the law to ensure landlords addressed housing conditions that affected the health of their tenants.

He had been an expert witness in many of the leading cases involving damp homes, for example *Nottingham City District Council v. Newton* [1974] 1WLR 923; *GLC v. London Borough of Tower Hamlets* (1983) 15 HLR 54 and *Birmingham DC v Kelly* (1985) 17 HLR 572 demonstrating that the Statutory Nuisance provisions of the Public Health Act 1936 (now the Environmental Protection Act 1990). In these cases, the tenants took their own action under s.99 of the 1936 Act (now s.82 of the 1990 Act).

He was also the expert in *Patel v Mehtab* [1980] 5 HLR 78, QBD, where the courts clearly accepted that an environmental health officer could be an expert and give expert evidence that the Court could not ignore.

In 1982, in *Legal Action*, David wrote an article, "Condensation", for lawyers and others involved in legal actions relating to condensation and mould, to "counter the suggestion often made . . . that increasing the heat and opening the windows is all that is necessary" to solve the problems. Unfortunately, that "lifestyle" defence was still being used 40 years later, so the first edition of this book was intended to try counter that. David also wrote an updated version of the 1982 article in 2023, and that article is referred to in several places in this edition.

Prof. David Ormandy

Contents

Series Preface

This series is ever expanding illustrating the issues relevant to environmental health. Although this subject is the first to have a second edition, illustrating increased concern with the topic.

The aim of the series remains as ever; to explore environmental health topics traditional or new and raise sometimes contentious issues in more detail than might be found in the usual environmental health texts. It is also a means whereby environmental health issues can be discussed with a wider audience in mind.

This series is an important part of the professional landscape, as is apparent from the titles published so far. Environmental health practitioners (EHPs) bring their expertise to a range of situations and are deployed differently but not always to the best effect so far as public health is concerned. All too often politicians both at the national and local levels are unaware of what is environmental health, and what practitioners do or how they work. It is common that practitioners have a "low profile" or are taken for granted. It is hoped that this series will be used as a means of highlighting the work of environmental health practitioners.

We want to encourage readers and practitioners, particularly those who might not have had work published previously, to submit proposals as we hope to be responsive to the needs of environmental and public health practitioners. I am particularly keen that this series is seen as an opportunity for first-time authors and as ever would urge students (whether at first- or second-degree level) to consider this an avenue for publishing findings from their research as some have already done. Why for example should the hard work that has gone into a dissertation or thesis lie unread on a library shelf? We can provide advice on turning a thesis into a book. Equally this series can be a way of extending a presentation, paper or training materials, so that these can reach a wider audience.

The series provides a route for practitioners to improve the profile of the profession as well as provide a source of information. It has the advantage of having a relatively quick turnaround from submission of the manuscript to publication and can be more up to date and immediate than a standard text-book or reference work.

It seems that EHPs have perhaps not been good at telling others about their work and contribution to public health and how they can help at an individual level. Perhaps that is due to concern that diminishing resources could lead to them being overwhelmed with demands. Yet to be considered a genuine profession and to develop professionally EHPs on the front line need to "get published", writing up their work of protecting public health. This series is a route for analysing actions and reporting on what worked in practice, what was successful, what wasn't and why. This can provide useful insights for others working in the field and also highlight policy issues of relevance to environmental health.

Contributing to this series should not be seen merely as an exercise in gathering CPD hours but as a useful method of reflection and an aid to career development, something that anyone who considers themselves a professional should do. I am pleased to be working with Routledge to provide this opportunity for practitioners.

As has been made clear it is not intended that this series takes a wholly "technical" approach but provides an opportunity to consider areas of practice in a different way, for example looking at the social and political aspects of environmental health in addition to a more discursive approach on specialist areas.

Our hope remains that this is a dynamic series, providing a forum for new ideas and debate on environmental health topics. If readers have any ideas for titles in the series, please do not be afraid to submit them to me as series editor via the e-mail addresses below.

"Environmental health" can be taken to mean different things in different countries around the World and so we welcome suggestions from a range of professions doing "environmental health" work or policy development. EHPs may be a key part of the public health workforce wherever they practise, but there are also many other practitioners working to safeguard public and environmental health. It is hoped that this series will enable a wider range of practitioners and others with a professional interest to access information and also to write about issues relevant to them.

Forthcoming monographs are likely to cover such topics as air pollution and monitoring private water supplies and even on ethics and environmental health. We are in contact with colleagues around the world, encouraging them to submit proposals. That does not mean we have no need of further suggestions, quite the contrary, so I hope readers with ideas for a monograph will get in touch via Ed.Needle@tandf.co.uk.

Stephen Battersby, MBE PhD, FCIEH, FRSPH

Series Editor

Introduction

As in the first edition, this book provides a definition of dampness in each of its forms and details the various potential sources, causes that can result in damage to the building, and, importantly, the threats to the health of the occupiers. It is practical, providing an outline of the possible solutions and looking at aspects of building design and construction that can reduce or avoid the risk of dampness.

It also discusses why dampness is a risk to the health of occupiers regardless of tenure and so justifies the need to protect health by reducing or removing it. This also leads to legal liabilities arising from dampness in rented dwellings

Co-authored by a medical doctor and environmental health practitioners with a combined experience of over 50 years, this book includes:

- Explanations and justifications for why dampness is important and why remedial action must be taken.
- Up-to-date information on the causes, effects, and remedies of dampness in the housing environments.
- Consideration of the legal liabilities and remedies for damp housing.

Dampness in Dwellings is a pivotal resource for active professionals in housing, medical, and legal sectors. The lead author of the first edition was the late Professor David Ormandy, a public and environmental health advisor, housing and health consultant, researcher, and author based at the University of Warwick, UK. He died in April 2024, a few months after the proposal for the second edition had been submitted to the publishers.

The decision to proceed was taken in his memory, and the reasons are in the dedication.

Table of Cases

Foreword

Dampness in the dwelling is perhaps the most important reason people can be dissatisfied with what should be their home.[1] If they are tenants, they will complain to their landlord, or if they are owner-occupiers, they will worry about what to do. If the landlord does not respond promptly or at all, tenants may well complain to the local environmental health practitioner or advice agency. There are several structural problems that can result in dampness, and that dampness can cause damage and distress and can also have a negative impact on physical and social health and wellbeing.

Yet as the housing ombudsman said in the Spotlight Report when it came to housing providers, "Our call for evidence revealed an immense frustration and sense of unfairness at the information residents are sometimes provided by landlords about issues like condensation and mould. This reoccurred so often it is appropriate to call it systemic," and all too often the occupiers and their "lifestyle" were blamed for the dampness.[2] That this is wrong and costly for social landlords has been illustrated in 2024 by the Housing Ombudsman finding severe maladministration, ordering a local authority landlord (LB Lewisham) to pay nearly £40,000 in compensation for three cases where the landlord failed to investigate damp and mould properly and simply "attributed a lot of the problems to the resident's lifestyle".[3] The Housing Ombudsman also made four findings of severe maladministration against the Guinness Partnership, with vulnerable children living in damp and mould for years and another resident with fungi growing in one of the bedrooms, and ordered total compensation of more than £15,000.[4]

The UK Health Security Agency study has also concluded "that ethnic and minority and disadvantaged groups were disproportionately impacted by damp and mould reflecting a persistent environmental health inequity".[5]

This monograph is intended to provide a straightforward explanation of the possible causes and potential threats to health from dampness. It follows the order of the first edition plus a chapter on the law. So it not only examines possible remedies for the dampness but includes legal remedies. It is largely non-technical and written for every tenant and owner-occupier whose home is damp, as well as advice agency workers and housing and legal professionals. The information should also help social landlords improve their administration when it comes to dampness. For environmental health practitioners, it will be of use to those who require their knowledge updating.

It aims to highlight the link between problems with the dwelling design and structure and the interference with the occupiers' efforts to create a healthy home environment.

Notes

1 The WHO LAREs study (The Large Analysis and Review of European housing and health Status project) distinguished "home" and "dwelling"

   Home is the social, cultural and economic structure created by the household representing a refuge from the outside world enabling development of a sense of identify and attachment

   Dwelling is the physical structure providing shelter and the necessary space, facilities and amenities for the household, it should pose no threat to health and safety of occupiers or visitors, and enable the establishment of the "home". See Ormandy D, Ed. (2009) *Housing and health in Europe – the WHO LARES project*, Routledge, Abingdon, Oxon.

2 Housing Ombudsman Service (2021) *Spotlight on: Damp and mould, it's not lifestyle*, Housing Ombudsman Service, Liverpool. Available at https://www.housing-ombudsman.org.uk/wp-content/uploads/2021/10/Spotlight-report-Damp-and-mould-final.pdf

3 https://www.housing-ombudsman.org.uk/2024/03/14/lewisham-councils-failings-in-damp-and-mould-complaints/

4 https://www.housing-ombudsman.org.uk/2024/03/19/guinness-partnership-failings/

5 See https://research.ukhsa.gov.uk/our-research/damp-and-mould/

1 Introduction and Background

Dampness in dwellings has long been a problem but not always as a health issue. One form of dampness was for a long time not given the serious attention it warranted outside of the environmental health profession. In 1983, the then Institution of Environmental Health Officers published "Background Notes on Condensation in 1983" reflecting concern that landlords and their agents were blaming tenants' lifestyles without proper investigation of the causes. It should also be remembered that circular advice from the government in 1989[1] made it clear that condensation was just one form of dampness, yet many who should have known better would say to tenants, "it's not dampness it's condensation". As this circular related to interpretation and enforcement by local authorities, it is suspected that many landlords, particularly in the social rented sector, were either unaware of this or ignored it.

Dampness leads to decay and deterioration of the materials used in the construction of dwellings, but the health risks have now been more widely appreciated. This concern has been particularly heightened since the coroner's report on the death of Awaab Ishak in Rochdale. Here it was concluded that the presence of condensation mould had contributed to Awaab's death. Up until now, although mould might be more obvious to occupiers and could cause ill-health, it was not thought that dampness could contribute to the most severe health outcomes, unlike other hazards such as falls or fire.

The coroner's report also suggested "there was no evidence that up-to-date relevant health information pertinent to the risk of damp and mould was easily accessible to the housing sector". This book attempts to rectify this at least in part, although, as the HHSRS Operating Guidance suggests, it is for users to keep up to date with current research[2] and perhaps that is what professional bodies or the government should address.

When the HHSRS was developed in 2004, there was a recommendation that the means to keep users informed of developments was made but rejected. Since then, various attempts have been made to provide such resources for all hazards, and the latest can be found on the Healthier Homes Partnership website.[3]

DOI: 10.1201/9781003586852-1

The Rochdale case was, regrettably, not unique. The coroner of Nottingham City and Nottinghamshire, Neema Sharma, has written to the head of housing at the social landlord Mansfield DC about the death of one of its tenants, Jane Bennett.[4] After moving into her council home in October 2022, Mrs Bennett reported issues with mould. In May 2023, she was admitted to hospital with respiratory issues, and she died the following month, aged 52. A post-mortem examiner reported, "in my opinion death was due to acute infective exacerbation of interstitial lung disease. Mould in her house could have contributed for the development of interstitial lung disease". The coroner was

> *concerned that mould in Mrs. Bennett's property and other properties in that area owned and maintained by [Mansfield] may pose a risk that future deaths could occur. I ask for the aforesaid properties to be inspected for mould and action be taken to ensure any further exposure to mould by any tenant is minimised.*

It is not clear who the coroner was expecting to undertake such inspections, or indeed who was competent to do so, although hopefully this book will help in the future. As is made clear in Chapter 6, the local authority EHP could not act under Part 1 of the Housing Act 2004, as the local housing authority on whose behalf the EHP operates cannot act against itself. This is a legal lacuna that ought to be addressed in the Renters' Rights Bill but is not satisfactorily in the version introduced in September 2024. This is discussed in Chapter 6 along with what action is available to council tenants.

The concern at government level so far as England is concerned can be gauged by information published on the government website in 2023[5] and also in a letter to Social Housing Providers[6] drawing attention to that guidance. In November 2022, the Secretary of State Michael Gove had written to the Local Authority Chief Executive and council leaders,[7] considering it "necessary and urgent to ensure that, as we go into a challenging winter, damp and mould issues are being addressed". He directed, under section 3(3) of the Housing Act 2004, that all local housing authorities in "carrying out their duty to review housing conditions in their area must:

- Have particular regard to high scoring (bands D and E) Category 2 damp and mould hazards, as outlined in the guidance Housing Health and Safety Rating system (HHSRS) enforcement guidance: housing conditions;[8]
- Supply the department with an assessment of damp and mould issues affecting privately rented properties in your area, including the prevalence of Category 1 and 2 damp and mould hazards; and
- Supply the department with an assessment of action you have identified that may need to be taken in relation to damp and mould issues affecting privately rented properties in your area."

The outcome of this direction is unknown at this time.

According to Clark et al.,[9] referring to data from the English Housing Survey (EHS), around 3–4% of homes have damp and/or mould in England (for 2022 the number was 1.027 million dwellings with any damp[10]); however, self-reported information from the Energy Follow Up Survey indicates that damp and/or mould may be present in as many as one in four homes (25%).[11] In 2022, there were an estimated 25.2 million residential dwellings in England, which implies 6.3 million dwellings with some dampness.

As of 2023, the Building Research Establishment (BRE) estimated that 64,708 dwellings had the greatest threats to health from damp and mould[12] – that is the worst case which would represent around 6% of all dwellings assessed with damp in the EHS.

Citizens Advice research[13] found in a survey of 2,000 private renters that 70% had experienced cold, damp, or mould in a property they had rented. It was also reported that there are currently 2.7 million renting households experiencing damp mould or cold, including 1.6 million children. The survey found 42% said that damp mould and excessive cold had increased their energy bills, 40% felt stressed as a result of damp mould and excessive cold, 35% said it made them feel anxious, 12% said it made them spend less time at home, and 8% said it had made their respiratory illness worse.

According to government figures, in 2018 in Wales around 7% of dwellings had damp/mould. Around 10% of 2.67 million dwellings (267,000 in 2021) in Scotland had damp and mould. In Northern Ireland, it is not clear in 2023 how many out of a stock of 828,829 had damp and mould (in 2016, the House Condition Survey indicated 9,300 dwellings were classified as unfit due to dampness out of a stock of 780,000 dwellings). As an interesting comparison, 26.1% of Australian homes are said to have dampness problems.[14] It is clear that damp and mould is an issue in a number of Australian states.[15] It is also a problem in New Zealand, where in 2018, 21.5% of all dwellings were sometimes or always damp.[16] Mouldy homes followed a similar geographical distribution to damp homes, but the 2018 Census showed only 16.9% of dwellings had mould.

The BRE suggests in England the cost to the NHS of a home with a Category 1 hazard of damp and mould is £521 per year, but remedying all Category 1 hazards (the worst) for damp and mould would save the NHS £33.7 million a year with a payback period of 7.5 years if all were mitigated.[12,17] The costs to society as a whole (exported costs) are far greater, and this ignores those dwellings that have some damp and mould, but the threats to health are seen as not as great – even though they are greater than in a dry dwelling.

There are different causes of dampness that can affect dwellings and a series of names to describe them – penetrating, traumatic, rising, condensation, and so on. Whatever the cause, dampness in dwellings has long been recognized as an important issue. In England and Wales, "freedom from damp" was always one of the criteria in the standard of fitness,[18] and the dwelling house "shall be deemed to be unfit for human habitation if and only if it is so far defective" (by the presence of dampness) "that it is not reasonably suitable for occupation in that condition", although the situation is somewhat different in Scotland.[19] This standard in England and Wales was replaced by the risk assessment approach in the Housing Health and Safety Rating System (HHSRS), and one reason was the variation in interpretation as to what was "reasonably suitable for occupation".

The aim of the first edition of this monograph was to summarize the structural and design matters that can lead to dampness, describe some of the structural problems that result from or are encouraged by dampness, and discuss and highlight the potential threats to health from dampness. As well as looking at the problems associated with dampness, the aim was and remains to offer some possible solutions. This will include looking at aspects of building design and construction that can reduce or avoid the risk of dampness and the remedies that can solve problems.

The original idea was to provide information about dampness without too much technical jargon. Unlike many publications, as has been made clear, there is a stress on and explanation of why dampness is a risk to the health of occupiers, highlighting the need to avoid or remedy dampness to protect health. Since the original publication, it has become apparent that there is

Figure 1.1 The purpose-built blocks of flats (single-level dwellings). It should be noted that both the orientation and height will have an effect on the internal environment, including temperature, and "boldly expressed structural members" can also be "cold bridges", discussed later.

a need, and it is appropriate to include a little more technical information, although again it is hoped that it has been written in a way that a non-technical person might understand.

As in many areas of housing and health, there is limited up-to-date information that can be relied on by surveyors, local authority officers, building owners, and lawyers. That said, this monograph still aims to provide a resource both for those involved in the technical aspects of dampness and dwellings and for those who want relatively non-technical information.

It provides some more detailed information on assessing dampness for those whose role is in determining the risks to health. As a result of comments on the first edition, a chapter is now included that also puts the issue of dampness in a legal context. This means that the book also sets out in brief the legal liabilities of landlords arising from dampness in dwellings. This can be a complex area of law, but it is hoped that it is readable and accessible for the non-lawyer and also is relevant as the law in England has been amended in response to the Awaab Ishak case.

This book perhaps sums up what environmental health is and what environmental health practitioners are about, relating technical knowledge to understanding of health and health impacts and putting it to use while also appreciating and using the relevant legal framework to resolve problems.

Notes and References

1 DoE Circular 6/90 – Circular issued following amendment to the standard of fitness in the Housing Act 1985 by the Local Government and Housing Act 1989, HMSO.

2 ODPM (2006) *Housing Health and Safety Rating System, Operating Guidance. Guidance about inspections and assessment of hazards given under s.9 at para 2.18 in the context of the 'ideal' "this will change, and it is the responsibility of those using the HHSRS to keep up-to-date on what is the ideal" and at para 4.09 "It is imperative that users of the Rating System keep up to date with published research and other relevant information which can be used to supplement that given in the Hazard Profiles (Annex D) and which may influence their judgment as to likelihood and/ or spread of harms."*

3 See https://www.healthierhousing.co.uk/home

4 *Jane Bennett – prevention of future deaths report* (Courts and Tribunals Judiciary, 8 Dec 2023). Available at https://www.judiciary.uk/prevention-of-future-death-reports/jane-bennett-prevention-of-future-deaths-report/

5 https://www.gov.uk/government/publications/damp-and-mould-understanding-and-addressing-the-health-risks-for-rented-housing-providers/understanding-and-addressing-the-health-risks-of-damp-and-mould-in-the-home–2

6 Letter from Michael Gove Secretary of State to providers of social housing 11 September 2023 at https://assets.publishing.service.gov.uk/

media/64feeace1886eb000d97707f/To_Social_Housing_Providers_from_Secretary_of_State_DLUHC.pdf

7 "Housing Standards in Rented Properties in England" – letter to Local Authority Chief Executive and council leaders at https://assets.publishing.service.gov.uk/government/uploads/system/uploads/attachment_data/file/1118877/SoS_letter_to_local_authority_chief_executive_and_council_leaders.pdf

8 ODPM (2006) *HHSRS Enforcement Guidance, Housing Act 2004 Part 1 housing conditions*. https://www.gov.uk/government/publications/housing-health-and-safety-rating-system-enforcement-guidance-housing-conditions

9 Clark SN, Lam HCY, et al. (2023) The burden of respiratory disease from formaldehyde, damp and mould in English housing. *Environments*. 10(8):136. https://doi.org/10.3390/environments10080136

10 DLUHC (2023) *2022–23 English housing survey headline report, Annex Table 4.5: Damp problems, 1996 to 2022*. Available at https://www.gov.uk/government/statistics/annex-tables-for-english-housing-survey-headline-report-2022-to-2023

11 BEIS. (2021) *Energy follow up survey: Thermal comfort, damp and ventilation: Final report*, BEIS, London, UK, 2021. Available at https://assets.publishing.service.gov.uk/government/uploads/system/uploads/attachment_data/file/1018726/efus-thermal.pdf

12 Garret H, Mackay M, et al. (2023) *The cost of ignoring poor housing*, Building Research Establishment, Gartson, Herts. Available at https://files.bregroup.com/corporate/BRE_the_Cost_of_ignoring_Poor_Housing_Report_Web.pdf

13 Citizens Advice. (2023) *Damp, cold and full of mould: The reality of housing in the private rented sector* (See https://assets.ctfassets.net/mfz4nbgura3g/UYinLQM79sdfwz52aDPkh/a067dd40fe0584e5e6242e50e564726b/Damp__20cold_20and_20full_20of_20mould_20_1_.pdf).

14 Knibbs LK, Woldeyohannes S, et al. (2018) Damp housing, gas stoves, and the burden of childhood asthma in Australia. *Med J Aust*. 208(7):299–302. https://doi.org/10.5694/mja17.00469

15 Tenants Victoria. (2023) *The mould report – a rented snapshot*. Available at https://tenantsvic.org.au/articles/files/reports/The-Mould-Report-A-Renter-Snapshot-May-2023.pdf

16 See https://www.infometrics.co.nz/article/2020-10-too-many-houses-are-mouldy-damp-and-cold

17 It is worth noting that such an assessment could not be made without the HHSRS methodology and the development of which can be read at the HHSRS archive at https://www.sabattersby.co.uk/

18 See s.4 Housing Act 1957 to 604 Housing Act 1985 as amended by the Local Government and Housing Act 1989.

19 It should be noted that the Scottish Tolerable Standard in s.86 of the Housing (Scotland) Act 1987 (as amended) is met if among other things the house is substantially free from "rising or penetrating damp" and excludes condensation directly although it might be argued that the requirement that it "has satisfactory provision for natural and artificial lighting, for ventilation and for heating" and "has satisfactory thermal insulation" might pick this up but there is no guarantee.

2 What Is Dampness?

There is no internationally recognized or agreed definition of what constitutes "dampness". So here, we explain what we mean by the term "dampness". Throughout the book, we try to make clear that condensation is just one form of dampness. Too often, people say (or have said) that it is not dampness, but it is condensation – this is plainly wrong.

Water (moisture) is naturally present in many of the materials used in the construction of buildings. Provided that the water or moisture stays within certain limits (depending on the particular material), it will not cause any problems. If the moisture exceeds the upper limit for that material, problems (deterioration) will occur, and this is what is referred to here as "dampness".

Water vapour, a gas, is always present in the atmosphere. Below certain upper limits, it will cause no problems. Above those limits, there can be problems for both the building materials and health, and these problems (threats) can occur before there are any signs of visible condensation.

Moisture in Building Materials (Construction Dampness)

Moisture is held in building materials in several ways. Water is combined with some building materials such as concrete and plaster. In the construction of a (say) two-storey, two-bedroom house, this can be in excess of 2,000 litres (depending on the form of construction). Most of this will dry out, but it may take at least 12 months and maybe more, and what is left is chemically combined with the material (and will not cause any problems). It is worth noting that until 60 or more years ago, new houses were not decorated for at least 6 months after completion and then only with porous finishes such as "distemper" or "breathable" emulsion paints. Current construction methods are drier and use less water.

Porous building materials, including plaster, concrete, bricks, and timber, will exchange moisture with the adjacent atmosphere as a result of vapour

DOI: 10.1201/9781003586852-2

pressure. Vapour pressure attempts to keep a balance in the moisture levels, as moisture from high-pressure areas will force moisture into low-pressure areas.[1] Normally, this vapour pressure will not disrupt the natural moisture levels of building materials, and as the porous materials are never "truly dry" (i.e. there is always some moisture present as well as any chemically bonded water), their normal state is usually referred to as "air-dry".

Examples of the moisture levels for some air-dry materials in a relatively moist, but not "damp" atmosphere, are – common bricks (not very dense engineering bricks), which will be between 1.5% and 2.5%, plaster around 1.0%, and timber around 11.0% (depending on the type of timber).

Hygroscopic Salts

Porous materials, such as brick, plaster, and concrete, can become contaminated by inorganic hygroscopic salts. These salts have an affinity for moisture and will absorb moisture from the air, disrupting the balance, making the material visibly damp. There are two potential sources. Salt staining can also occur on both the outside and inside of walls, the salts being drawn from the soil by rising dampness and left as the moisture evaporates.

One is where there is a solid fuel heating unit ischarging into an unlined flue, where the flue gases cool and condense, passing salts into the chimney breast. When dampness has been rising in a wall for some time, the soluble salts contained in the ground become concentrated where the water evaporates. This can happen after the rising dampness has been dealt with such as by the insertion of a new damp proof course. These deposits of salts can absorb water directly from the air to such an extent that the wall can become visibly wet. This dampness effect is usually referred to as "hygroscopic" dampness. It can lead to the assumption that the rising dampness (discussed below) has not been remedied or can be confused with condensation.

Rising Dampness

Porous materials are riddled with very fine hair-line pores, and these pores are so fine that the surface tension of water will become strong enough to draw the water upwards against gravity (the same effect as a wick or blotting paper). This effect, capillary attraction, means that materials such as bricks and concrete will "draw" moisture out of the ground to heights of about 1 m above ground level. As noted previously, the moisture can contain hygroscopic salts which, as well as the damaging effect of the moisture, will further damage the walls' structure. Rising dampness can also have a dramatic effect on a solid floor that is in direct contact with soil.

It is to prevent rising dampness that damp-proof courses (DPCs) and damp-proof membranes (DPMs) are or should be incorporated into walls and floors in direct contact with the ground. In older houses (those built over

100 years old), there was often a dado, wood panelling from floor level to just less than 1 m (approximately 3 feet) above. The dado was a decorative finish that hid the damage caused by rising dampness, and did not cure the problem. The DPC in the wall should be 150mm (6 ins) above the external ground level. If the DPC in a wall is "bridged" that is, the external ground is at the same level or higher than the DPC, then the effectiveness of the DPC will be compromised, and can result in a damp wall.

Penetrating Dampness

Holes and gaps in the external fabric, the walls and roofs that should protect the interior, will allow water through that protection. Such holes and gaps may be a result of despair and lack of maintenance, or ineffective weatherproofing at the time of construction or refitting. Some gaps can be obvious, such as a slipped roof slate or tile. Often, particularly in older properties, walls will have a coating of external render (a waterproof concrete skim) to protect walls constructed of brickwork. Cracks in such render will draw water in when it rains, and as that water will not evaporate outwards, it will soak into the wall to affect the internal surfaces.

Of particular importance is the prevention of water penetration at the joints around window and door openings. These must be properly and completely sealed, and, for windows, there should be a sill at the base of the opening to throw any water running down the glazing safely away from the wall below.

Brickwork is porous and rain can penetrate to a certain shallow level. It will normally evaporate outwards, but over time, if the water freezes and expands, it can cause the external face of the brickwork and pointing to be pushed off (spalled). This freeze/thaw cycle can lead to deterioration of the masonry (brickwork). If this is not addressed in older properties, rainwater can penetrate deeper into the structure, which in solid walls can lead to internal dampness.

Traumatic Dampness

This is when a water pipe or tank leaks or bursts, or as a result of a leak to drainage or waste pipe serving a water closet (WC), sink, bath, or shower. The effect can be slow, where the leak (often at a poorly made joint) is relatively minor, but this will be enough to cause a problem over a period. Or, it can be dramatic, such as when a water tank bursts.

Problems can also occur if the water in an uninsulated pipe or water tank freezes. Frozen water (ice) expands and can break a joint or burst a pipe, and a leak resulting from this freezing will occur when the ice melts. As water pipes and tanks are generally hidden from sight, any traumatic dampness can be difficult to trace and remedy.

Moisture in the Atmosphere

The amount of water vapour (moisture) a given volume of air can hold depends on the temperature of that air – that is, the amount of moisture (a gas) that air can carry at any barometric pressure depends on the temperature of that air; the higher the air temperature, the more water vapour it can hold. The term "Relative Humidity" (RH) is the ratio between the amount of water vapour held by a volume of air and the maximum amount that volume of air is capable of holding at that temperature.[2] Thus, if a heating system heats the air up quickly but the structure does not and only responds slowly, remaining cool, the risk of condensation is increased. Condensation does not always occur where the moisture is generated, as moisture-laden air tends to move from relatively warm to relatively cold areas and from relatively wet to dry area, as moist air (containing more water vapour) has a higher pressure than areas with a lower vapour pressure. Thus, condensation can occur in rooms that may not be used, or are less well heated or are unheated.

Ideally, the RH within a dwelling should be between 30% and 70%; any lower and it will start to feel uncomfortable (dry), and any higher and it will start to cause problems, such as mould growth. However, within a room (and a dwelling), there will be air temperature gradients, both horizontally and verti-cally; the temperature is highest close to heat sources and lowest next to cold (or cooler) surfaces. This means that the RH will differ from point to point, although the amount of water vapour may not vary.

If the RH within the dwelling persistently exceeds 70%, then "damp" problems will occur, although, initially, there will not be any visible signs of dampness. Visible dampness, termed "condensation", occurs when the air becomes saturated (i.e. the RH reaches or exceeds 100%). This is known as the dew-point. When the air is cooled, it can hold less water vapour, and when it reaches dew-point, any further cooling means the water vapour begins to condense to form liquid water. Condensation is usually visible on cold sur-faces that have reduced the air temperature, such as window glazing and cold spots on walls where heat has been transmitted from inside to the exterior.

Dehumidifiers can reduce the RH but will not address the underly-ing causes. They will also increase energy consumption so they are not a long-term solution.

However, condensation can also occur within the structure and so may not be visible. This happens when moist air passes through a material such as plaster (by vapour pressure) and reaches a cooler non-porous material, reducing the temperature and increasing the RH beyond the dew-point. This is termed "interstitial condensation" and may happen in timber-framed build-ings (where there is a timber frame, hidden from the outside by brickwork, and from the inside by plaster boarding), where walls have been dry lined to improve the insulation (an internal lining of plaster board on lathes) or where there is a flat roof with no or an inadequate vapour barrier.

As well as interstitial condensation, condensation can occur when moist air from the living areas is able to reach the roof space (loft) under a pitched roof and in the sub-floor space (the space below a suspended timber floor).

The production of water vapour in an occupied dwelling is a natural result of domestic and biological functions and activities. There are various estimates of the amount of moisture generated by a four-person household (see Table 2.1). A realistic estimate is that such a household can generate 30–40 litres of moisture per week just by breathing; add to that 15–20 litres a week by cooking, showering, and bathing. If the laundry is dried indoors (in inclement weather), this could add a further 35 litres a week. This means that such a four-person household will generate anywhere between 45 and 95 litres of water vapour a week purely from normal domestic and biological functions and activities.

In addition, washing floors would add another 200 g/day and moderate manual work 300 g/hr per person.

The British Standard estimates that an average moisture production rate for a family with teenage children, indoor drying of laundry, and so on, with four occupants would be 14 kg/day (14 litres). It should also be noted that the moisture production rates of fuels differ too – for example, natural gas will release 150 g/kWh. This should not be a problem for a gas fire as the water vapour should pass outside directly, either via a balanced flue or lined flue. Portable LPG will produce 100 g/kWh (the same a paraffin), but that will all be released into the dwelling. Using electricity does not release any water vapour.

There will be variations between different households, but these will be surprisingly small. Even a single-person household could generate between 5 and 10.0 litres over 24 hours and a two-person household at least 7 litres. These figures are only averages, and clearly some households will quite reasonably generate more water vapour. What is clear is that the dwelling should be designed to deal safely with such quantities of moisture. Table 2.1 indicates that as 1 litre of water weighs 1 kg, most households these days will generate more. The actual amounts emitted will depend on the size and composition of the household and the amount of time spent in the dwelling; for example, a household of two older people would spend more time indoors than a household consisting of a working couple. A household including young children may also spend a lot of time indoors. While a dwelling should be of a size and layout to suit the possible occupying household, this will not always be the case as occupation changes over the years (households moving, children being born, and later moving out). All this means that dwellings should be capable of being occupied by a spectrum of households without any condensation problems.

While the behaviour of occupiers will have some effect on the amount of moisture generated within the dwelling, this is unlikely to be a major contributor to high RH or condensation (see Table 2.1). Unfortunately, it is all too easy for building managers and landlords of rented dwellings to lay the blame at the door of the occupiers rather than solve why the dwelling does not meet the basic principle of providing living accommodation able to cope with occupiers. Crowding

Table 2.1 Typical moisture generation for household activities

Household activity	Moisture generation rates
People	
asleep	40g/hr- person
seated	70g/hr -person
standing/housework	90g/hr - person
Cooking	
electricity	2000g/day
gas	3000g/day
Dishwashing	400g/day
Bathing/washing	200g/day - person
Shower	600g/shower
Washing clothes	500g/day
Drying clothes indoors	1500g/day
Plants	20g/day - plant

Source: BS5250:2021 Management of moisture in buildings – Code of practice

can mean excess moisture, but this is a different problem. Expecting occupiers to change their lifestyle to reduce moisture production is expecting occupiers to compensate for problems attributable to defects including lack of maintenance, poor design, and/or construction of the dwelling.

Flooding

Two factors have increased the likelihood of dwellings being flooded. One is a result of changes in weather patterns that now mean there are more heavy rain storms and flash floods as the result of global heating, and another factor is the greater use of potential flood plains for housing developments.

UK houses were not, and are not yet, designed to avoid flood water entering. Flooding, even by a small amount, has a dramatic impact and requires extensive remedial action after the water has subsided, particularly as flood water will be contaminated, usually by sewage. This has been the subject of work by the Building Research Establishment (BRE).[3] Flooding is discussed further in Chapter 5.

Notes and References

1 Which is one reason why "opening the windows" is unlikely to be the answer to a condensation problem.
2 To understand this, think of a winter's day when the external air might be at 3°C it could be at 90% RH, whereas within the dwelling at 21°C could be at 30% it might be thought the external air contains more moisture but that would be wrong, the internal air contains more moisture.
3 https://bregroup.com/about/science-park/flood-resilient-repair-house

3 Sources of Dampness and the Potential Effect on the Structure

In this chapter we set out a summary of potential sources of dampness and the possible consequential effects. It cannot be considered comprehensive as so much depends on the form of construction and modern forms of construction will be very different from traditional methods. Full and more detailed information and consideration can be found in publications on construction, and the functions of structural elements. Some of these sources are listed in the Further Reading list at the end of the book but BS 5250:2021 provides some useful information and recommendations on the management of moisture in buildings using an integrated and pragmatic approach and includes some diagrams that readers might find useful.

Roofs

Pitched roofs are intended to direct water safely away to eaves gutters and rainwater goods. Pitched roofs are usually finished with tiles or slates (although some may be finished with metal, such as copper, or bitumastic felt).

The weakest point in any roof is where it is penetrated, such as by chimneys, pipes, and dormer windows. The weak point is the junction between the roof covering (tiles or slates) and the chimney, pipe, or dormer. Chimneys in older houses are usually made of brickwork, and cement or a malleable material (such as lead) is used to provide a water-proof seal, "flashing". This flashing, or the adjacent brickwork, may deteriorate allowing water to enter the roof space and dwelling, often causing dampness to the plastered surface of the chimney breast within the living area (See Figures A1.2 & A2.3).

So-called flat roofs (that is, roofs with a fall of less than 11 degrees) should be designed to prevent water from collecting (ponding), which can result in damage not readily visible and may therefore go undetected for some time.

Cold and warm deck flat roofs can be a problem particularly the former which are less common these days. For a cold deck roof, the insulation is placed immediately above the ceiling and between the joists. If the void above the insulation is not fully ventilated, then moisture can condense within the space. There should also be a vapour barrier on the warm side of

DOI: 10.1201/9781003586852-3

the insulation. Without these measures, the condensation within the structure can lead to damp patches on the ceiling which could be construed as penetrating damp as the result of disrepair to the roof covering. In a warm deck roof the insulation is placed above the roof decking that is fixed on top of the joists. This leads to the void in the roof structure being the same or similar as the room temperature reducing the risk of condensation. The vapour check should be installed again on the warm side of the insulation (under it), but the roof should not be ventilated as this allows cold air to enter and increase the risk of condensation.

Slipped slates or tiles, or damaged covering to a flat roof will allow water penetration into the roof structure, and that water may damage roof timbers or ceiling plaster to upper floors.

Rainwater Goods

These include eaves gutters intended to safely collect rainwater from roofs, and downpipes to collect the water from eaves gutters and take it safely away from the building. If these become blocked, water will overflow and can damage the adjacent wall, washing out the mortar of brickwork joints and the surface of bricks. The amount of water may result in penetration through the wall into the interior.

Where it is likely for snow to collect on roofs, it can slip as it starts to melt, and can result in too much water for the eaves gutters to cope, or the weight may over-power the eaves gutters causing them to overflow or even collapse under the weight. Where snow is regularly expected in winter periods, guarding should be fitted to retain the snow until it melts safely. Rainwater goods are vulnerable to blocking caused by freezing during cold periods. Ice expands as water freezes and can cause damage to downpipes.

Brickwork

Moisture in the soil will contain soluble salts, including salts of nitrate and chloride. These will be held in solution and, through rising dampness, into the structure of walls and solid floor. As the moisture evaporates, the salts will be left behind. Although the amounts are relatively minute, the salts deposited, over many years, will have some effects.

Brickwork in older houses without a damp-proof course will suffer with rising dampness. The salts deposited in the brickwork will initially leave a stain at the height reached by the moisture. This stain is, perhaps unsightly, but, in itself, not really a structural but a decorative problem. However, the dampness will damage the integrity of the bricks, and the outer surface will, eventually, loosen and spall (detach and fall-off). The salts in moisture from the soil are hygroscopic, absorb moisture from the air, and can, ultimately, have damaging effects.

Door and Window Openings

These are weak points in any external walls. The joints between the frames should be water-tight to prevent penetration. The sill to windows should slope away from the wall below and should incorporate a groove on the underside so that any water running off the window and frame drips away from the wall.

There should be a weather bar (or drip moulding) with a capillary (drip) groove fitted to the base of external doors (or a threshold bar where the closed-door fits) to prevent the penetration of water between the door and the internal floor surface behind.

Drip grooves to the head of the door and window openings divert water running down the wall away from the door and window openings. This applies also to external window cills. If the rainwater can run down the glazing on to the cill, without such an arrangement (so that the water drips down away from the wall), the water can be absorbed by the wall. If the pointing has perished or there is an open joint between wall and window frame, this can lead to a damp wall internally under the window.

External Drainage and Waste Pipes

These are the pipes taking soil and wastewater from facilities (such as baths, showers, wash hand basins, and water closets) from within the dwelling and discharging it into the drains and sewers, or sumps or septic tanks. Such pipe-work can become damaged or blocked, allowing the contents to spill onto adjacent wall surfaces, again damaging the structure. The water in these external pipes can freeze in extremely cold weather, but wastewater is less likely to suffer due to its higher temperature, but it is still possible. As the ice expands, it can, like rainwater downpipes, damage joints.

Solid Floors

Similar to brickwork in older houses, solid (concrete) floors without a damp-proof membrane will be affected by rising dampness. Most such floors will have a finishing surface on top of the concrete. While relatively protective from drying out and evaporation, the dampness will damage the material of the joint between the finish and the concrete so that the floor finish will become detached and loose.

Reinforced Concrete

Unprotected ferrous metals such as in reinforcement will rust (a reddish brown oxide of iron) when exposed to water and air. As concrete does not provide any protection in itself, rusting of the reinforcements can be a serious problem; as the metal rusts it expands, and the slow force generated can shatter the concrete.

It should be noted that concrete can be a good thermal conductor depending on its density and can be a "cold bridge" – discussed later in *Annex 1* (also see Figure 1.1).

In the past, pre-cast concrete houses have had damp problems after they have been insulated having previously been cold. Condensation occurred on the cold inner face of the external concrete panels and ran down to ground level and in some cases the build up of water soaked the bottom of the plasterboard linings and looked like rising damp.

Plasterwork

Rising dampness will affect internal plasterwork in older houses without an effective damp-proof course. Gradually, affected plaster will deteriorate, and any decorative surface may bubble and become detached from the underlying wall ("blown").

Internal plaster will also be affected by condensation, but this is less destructive to the structure, but will damage decorations and attract various forms of mould growth.

Ceilings (and walls) will be damaged by dampness penetrating through disrepair to roofs, by burst pipes, and leaking joints to sanitary ware, personal washing facilities, and storage tanks (in lofts). Penetration through roofs will be associated with rainfall. Traumatic dampness associated with facilities will be linked to their use; although this may be sudden and dramatic, the weight of the water may cause the plaster to collapse.

Although more of a nuisance when appearing in large numbers, rather than causing damage to the structure (or health) are plaster beetles (*Lathridiidae* spp.). These are small beetles (from 0.8mm to 3mm in length) that thrive in damp conditions (damp plaster) and can be found in new build housing (construction moisture), or if there is a problem of dampness in existing dwellings including high RH, traumatic dampness or the result of flooding and there is mould present on which the beetles can feed. They are more prevalent during the wetter months. Interstitial condensation can be a particular problem with mould occurring in hidden voids, providing a hidden food source for plaster beetles to live and breed.

Timber

Not all timber used in construction is visible, being an internal part of the structure. For example, "hidden" timbers can include floor joists supporting suspended ground floors, joists between ceilings and upper floors, joists supporting the ceiling to top floors, roof rafters, and timber within the structure of walls.

Unprotected timber (particularly softwood) used for windows and doors and their frames will be intermittently affected by dampness (such as rain and snow). This will cause warping and distortion. This can also be caused by a constantly damp environment. The result is that the windows and doors will stick and may not close and open easily or properly.

Damp-affected timber, with a moisture content of 18% or more, will be vulnerable to fungal attack. This is especially so for structural timbers (joists) to suspended ground floors where there is poor and inadequate ventilation. The most serious of timber-attacking fungi is generally known as "dry rot" (*Serpula lacrymans*). It produces a mass of mycelial (hair-like filaments) that spread through the timber, extruding enzymes that digest and weaken the timber. Dry rot is particularly virulent and can spread through brickwork and concrete. Although considered less serious but of importance is the so-called wet rot (*Choanephora cucurbitarum*). As well as damaging the structural integrity of timber, dry rot and other fungi produce spores, typically single-celled reproductive units. Such spores spread the fungus but also pose serious threats to health in addition to mould attributable to condensation.

Damp timber may also attract termites which can cause damage and weaken the timber as, unlike other wood boring insects discussed here, they go straight to the heartwood (although termites are rare in the UK, but widespread in parts of Europe and the USA). Dampness also has an impact in that most wood boring insects will only attack damp timber or are encouraged by dampness in the timber. Some insects have a preference for timber that has already been subject to fungal attack. Death watch beetle (*Xestobium rufovillosum*) only attacks hardwoods that have been affected by fungus. The common furniture beetle (*Anobium punctatum*) can attack dry wood unaffected by a fungus, but the life cycle is much shorter in timber affected by damp than dry. This means more larvae which do the damage to the wood. The small wood-boring weevil (*Euophryum confine*) is probably the second most common wood-boring insect seen in buildings across the UK. Again, it mainly attacks decaying and damp softwood and hardwood but can also damage plywood. Wood-borers' life cycles in cooler climates such as the UK mean that they are more active between April and August but can lie dormant for the rest of the year. The key to controlling insect infestations is controlling moisture and dampness in the dwelling[1].

Timber frame construction is a lightweight construction. The internal load-bearing leaf of the external cavity wall is formed of vertical timber studs. The outside of the studding is usually covered with plywood or other material as a "sheathing" to stiffen the frame. The internal face of the timber frame is covered with plasterboard with insulation between the studs. Weather protection is provided most often by an outer leaf of brickwork. This form of construction can be affected by "interstitial" condensation. This should be avoided if a vapour barrier is placed on the warm side of the insulation and under the plasterboard with joints in the vapour barrier material taped. Any holes for electrical socket boxes or similar installations need to be made with care to reduce the risk of condensation within the wall structure, which could affect the timbers.

Energy Efficiency

There are several factors that relate to the energy efficiency of a dwelling:

- Thermal insulation of the envelope (external walls, of windows, of roofs (or the ceiling to upper rooms), and of floors.
- Provision for space heating.
- Provision for ventilation.

Thermal Insulation

Older dwellings, constructed to meet earlier standards, are usually energy inefficient. The main structural factor is the thermal insulation provided by the structure (the fabric of the building). Brick-built walls were traditionally solid, and usually of 225 mm (9 inch) thickness. Such walls provide poor thermal insulation and poor protection from heavy rain. To improve protection from rain penetration, cavity walls were introduced; these had an inner skin of one brick thickness (about 115 mm, 4.5 inches), a gap (of around 50 mm, 2 inches), and an outer skin of one brick thickness. An unintentional gain was slightly improved thermal insulation provided by the air gap between the skins. Subsequently, requirements were introduced to increase the thermal insulation provided by external walls.

One method to increase the insulation given by external walls was the injection into the cavity of material such as polystyrene foam. Some forms of insulation to external walls can settle, reducing the insulation to upper sections of walls, and some can be affected by moisture, either from rising dampness or rain penetration. As water is a "good" conductor of heat, damp-affected insulation will be relatively ineffective, resulting in loss of energy and heat.

Windows in older dwellings often consist of a single pane of glass (unless upgraded) and will be of major heat loss component. Properly formed double or even triple glazing will reduce heat loss considerably, and so reduce the risk of condensation. Where double or triple glazing is not an option because of the type of window or planning restrictions, secondary glazing can help.

The ceilings to top floor rooms should be divided from the roof space (or the outer surface of flat roofs) with insulation (and a vapour barrier to prevent water vapour from passing into the roof space). Alternatively, the underside of the waterproofing outer surface of a pitched roof should be insulated; providing a useable loft space.

Suspended timber ground floors can lose heat into the under-floor space, and solid floors can transfer heat to the ground. In both cases, there should be a vapour barrier or damp-proof membrane, and, in the case of solid floors, an insulating layer.

Space Heating

The provision for space heating should be designed to distribute heat throughout the living area of the dwelling and be controllable and affordable. The type of heating should take into account the construction of the dwelling. Walls constructed on dense material will be slow to heat up and cool down, and will give off heat over a period. Heating systems that provide relatively instant heat, such as radiant and warm air systems, will heat the air and individuals quickly, but will be (very) slow to heat such structures. Lightweight inner surfaces will heat up quickly, but will not hold that heat.

Water heating should also be affordable and should be capable of heating the water to around 60°C; any hotter could cause scalding, and cooler may allow bacteria (such as *Legionella*) to multiply. Where a part of the provision for hot water includes storage tanks, these should be insulated to prevent heat loss from the water and so save energy. Cold water tanks and pipes should be insulated to protect against freezing.

Ideally, flueless gas or oil heaters should not be used in dwellings (although they are used where, for economic reasons, the occupying household has been disconnected from a mains gas or electricity supply and left with no alternative for space heating). The fuel for such heaters is not economic, and the heaters emit all the products from combustion as well as heat into the atmosphere within the dwelling (see Table 2.1 on the moisture emitted).

Ventilation

This is necessary to replenish the internal air and remove excess moisture. Means of ventilation should be controllable and not excessive and should help ensure relative humidity (RH) is kept at safe levels and heat loss is kept to a minimum.

In general, there should be around 0.5 air changes per hour, with increased localized extraction in specific areas, the kitchen, and bathroom, during periods of high moisture production (such as cooking and showering). Air changes above 0.5 for general parts of the dwelling will increase the heat loss. Windows do not need to be open wide, and most modern double-glazed units incorporate trickle vents, which should be sufficient in most rooms.

For the kitchen and bathroom, there should be extractor fans (or other mechanical ventilation system including, if possible, heat recovery) located and designed to take moisture-laden air out of the dwelling. Such fans can be fitted with humidity-sensitive switches, to automatically operate when moisture levels are high and with a manual control.

Problems associated with energy inefficiency as a result of an imbalance in the factors will be primarily high RH (i.e. above 70%) or condensation (visible moisture). These will enable mould and/or fungal spores (always present in the atmosphere) to germinate on surfaces. Where mould appears on wall or ceiling surfaces, it will not cause damage to the material, although it will damage decoration. The presence of high RH or condensation on timber will enable the spores to germinate and attack/infect the timber, ultimately weakening it.

In high-rise buildings there is also the stack effect to contend with. Warmer air is less dense than cooler air. As the warmer air rises, it creates a pressure difference, reducing the pressure in the base of the building, drawing cold air in through either open doors, windows, or other openings and leakage. This stack effect occurs mainly in the core of the building, such as stairway and lift shaft, and causes problems with energy loss as the result of the airflow. This has implications for energy bills.

Postscript

Although not strictly an effect on the structure, as has been made clear dampness in a rented dwelling can lead to both direct health effects (see *Chapter 4*) and to disrepair of parts of the structure, which in turn increases the landlord's liabilities and indeed opens the potential for litigation and legal action and this is now included in this edition at Chapter 6.

References and Notes

1 Parrett M. (2020) On the lookout for wood-boring insects. How do rotting timbers lead to infestations of wood-boring insects? *Property Journal*, May/June, pp. 48–50, RICS.

4   Health Effects of Dampness and Mould

Introduction

In the UK, the health effects of dampness and mould were highlighted by the death of two-year-old Awaab Ishak in Rochdale in 2020. Coroner Joanne Kearsley's report stated that Awaab "died as a result of a severe respiratory condition due to prolonged exposure to mould in his home environment. Action to treat and prevent the mould was not taken. His respiratory condition led to a respiratory arrest"[1]. This case, along with the widespread media coverage it received, including in medical journals[2], underscored that dampness and mould exposure in homes is an important public health issue requiring effective, coordinated action plans involving all stakeholders, including healthcare professionals and caregivers, to prevent similar tragedies.

The urgency of preventive action is heightened by the likelihood that the prevalence of dampness and mould in homes is set to increase in the current "cost of living" crisis context[3]. What's more, since the 2022 energy crisis, many European countries have recommended lowering indoor temperatures by 1 or 2 degrees Celsius, a measure that could lead to colder interior surfaces and an increased risk of condensation and mould growth. In addition, climate change, by increasing the frequency of flooding and heavy rainfall, especially in winter, could further exacerbate the problem of dampness and mould contamination in buildings[4]. Efforts to improve household energy efficiency could also increase the risk of dampness and mould growth if the "build tight, ventilate right" principle is not properly followed[5].

Dampness and Mould Growth

"Mould" refers to fungi that can grow on organic building materials as long as sufficient moisture is available. Of the at least 1.5 million fungal species, of which less than 100,000 have been described, the most found in damp homes include *Cladosporium sphaerospermum*, *Penicillium chrysogenum*, *Aspergillus versicolor*, *Alternaria alternata*, and *Stachybotrys chartarum*[6,7]. Moulds

DOI: 10.1201/9781003586852-4

produce spores, often suspended in air, which help them spread. They also synthesize chemicals, including mycotoxins and microbial volatile organic compounds (mVOCs), that may remain in spores or be released directly into the air[8].

Dampness and mould are common in homes worldwide, with prevalence rates from 10% to 50%, depending on the country or region[9,10]. The 2017 UK Energy Follow-Up Survey (EFUS) found that 39% of households with dependent children reported damp and mould, versus 23% of households without children. Lone-parent households were particularly at risk, with 48% experiencing damp and/or mould[11].

Increased indoor humidity is the primary factor promoting mould growth, while building materials and their moisture levels influence which fungal species proliferate and the types of mycotoxins and metabolites released into the indoor environment[12]. Besides indoor humidity, temperature, ventilation, and building materials also affect mould growth.

Damp-affected timber is vulnerable to fungal attack and infection. As spores of a wide range of fungi and moulds are always present in the atmosphere, damp timber provides an ideal medium for their germination and enables the fungus to become established. Wet rot (*Choanephora cucurbitarum*) and dry rot (*Serpula lacrymans*) are virulent timber-attacking fungi and will dramatically increase the spore content of the atmosphere within a dwelling, even though the source, the fungus, may not be visible. Other sources of spores are the moulds infecting damp plaster and other surfaces (both visible and hidden surfaces). These spores are encouraged by the presence of dampness in the material and by condensation as the result of high relative humidity (RH); indeed mould on walls is a symptom of high RH.

Mould growth is particularly common and severe in low-income communities, often due to insufficient insulation, inadequate heating and ventilation, and poor maintenance[13,14]. In newer low-energy buildings, increased airtightness combined with inadequate ventilation can raise indoor moisture levels and lead to surface condensation, often due to design flaws, poor installation, or lack of maintenance promoting mould growth.[12] Overcrowded dwellings, where more people reside than intended, further increase moisture levels, though this association is influenced by various confounding factors[15].

While the behaviour of occupants can influence the amount of moisture generated within a dwelling, it is unlikely to contribute significantly to high relative humidity (RH) or condensation. Moisture generation is primarily determined by household size, composition, and time spent indoors; even single-person households can produce 4.5–10 litres of moisture daily, with an additional 5–8 litres generated irregularly. Therefore, occupants should not be blamed for condensation or high RH; homes should be designed for their intended occupancy, and overcrowding should be avoided.

Health Effects Associated with Exposure to Damp and Mould in the Home

Public health organizations, including the WHO, recognize indoor mould growth as a significant health hazard.

Inhalation is considered to be the main route of human exposure to the substances produced by damp and mould in indoor environments. Moulds pose three main types of health risk: allergic, inflammatory, and infectious. The allergic risk is the most documented, particularly in the respiratory tract. The toxic and inflammatory risk is linked to exposure to mould components (fungal spores, β-glucans, and other components of the mycelium cell wall) and/or their metabolic products (mycotoxins, mVOCs). Microbial VOCs and β-glucans can exert pro-inflammatory responses through non–IgE-mediated mechanisms[16]. The risk of invasive mould infection is of greatest concern to people who are immunocompromised due to a medical condition or an immunosuppressive treatment[17,18].

Most studies that use observable signs of dampness such as visible mould, mould odour, moisture, or water damage as a proxy for mould exposure show increased associated health risks[19]. Quantifying mould exposure remains challenging. A lack of consensus on the method for assessing exposure to moulds has been highlighted for many years, particularly in reports by the IOM (2004) and the WHO (2009). This may explain the relatively large number of studies that have failed to demonstrate a direct association between quantitatively assessed mould exposure and health effects. Furthermore, the role of mould in these health effects is difficult to isolate from other dampness-related microorganisms, such as dust mites and bacteria, as well as non-biological particles and chemical pollutant[6,20].

The effects observed are either acute or chronic and of variable severity. It is not always easy to distinguish allergic effects from so-called non-allergic inflammatory effects despite similar symptoms[6]. Sensitization rates to fungi typically exceed 5% of the general population with higher rates among the atopic population[6,21]. The hygienist hypothesis, which favour a protective role for fungal metabolites in asthma and allergy, especially at early ages, is not supported by sufficiently robust scientific evidence. In fact, individuals that are sensitized to mould may develop allergic reactions when exposed to indoor moisture/mould damage. Therefore, to prevent sensitization as well as to protect against allergic reactions in the case of existing sensitization, in particular for patients with allergic asthma or allergic bronchopulmonary aspergillosis (ABPA) who are at risk of a worsening of lung function the mould infestation must be remediated[17,22].

Those more at risk of adverse health outcomes from damp and mould include pregnant women and their unborn babies, children, elderlies, and people with chronicle health conditions, in particular respiratory and cardiovascular, as well as immunocompromised persons. These more vulnerable people are also generally those who spend most of their time in their homes and are therefore more exposed to damp and mould[23].

Respiratory and Allergic Health Effects

The majority of studies assessing the health effects of inhalation exposure to fungi focus on asthma and allergic rhinitis. In 2004, a comprehensive review of the scientific literature by the Institute of Medicine found sufficient evidence for an association between damp environments and some upper respiratory tract symptoms like coughing, wheezing, and asthma symptoms in sensitive individuals[24]. In 2009, WHO expanded the IOM observed associations in an extensive review and guidelines on dampness and mould, which include asthma development, current asthma, dyspnoea, and respiratory infections[9]. In the last few years, additional epidemiologic research has been conducted and many meta-analyses and systematic reviews have been published confirming consistent positive associations, found in both allergic and non-allergic individuals, between indicators of dampness or mould in the home and various health outcomes: asthma development and exacerbation, dyspnoea, wheezing and cough, respiratory infections, bronchitis, hypersensitivity pneumonitis, allergic rhinitis and eczema[7,22,25,26,27,28].

The effects of exposure to mould on respiratory health have been demonstrated mainly for childhood asthma, with strong arguments suggesting causality. Children are more at risk because their respiratory system is still developing. Early-life exposure to moulds can cause recurrent irritation and immune activation in the respiratory tract, leading to prolonged inflammation and the development of inflammatory-related diseases like asthma and rhinitis. In a prospective birth cohort study, exposure to mould or dampness indicators during infancy increased the odds of asthma through adolescence[29]. Studies indicate that up to 14% of childhood asthma is attributable to living in a home with damp or mould[16].

A recent meta-analysis found that living in mouldy homes increases the risk of asthma in children by 53% according to 21 case-control studies, and by 15% according to 11 cohort studies[30]. Studies, particularly panel studies, also show that exposure to moulds aggravates asthma symptoms in children[7] and that infants exposed to mould odour or visible mould have an increased risk of rhinitis[28]. A systematic review and meta-analyses of epidemiological studies found that early life exposure to residential mould and dampness indoors has a small to moderate effect on the prevalence of respiratory tract infections (RTIs)[31].

When considering children at risk of poorer respiratory health due to poor-quality housing, it is also important to consider antenatal determinants of lung development and the environment in which the mother lives during pregnancy. Studies have shown that intra-uterine exposure to PM associated to outdoor air pollution had a negative impact on respiratory health in childhood, but evidence assessing the effects of intrauterine exposure to indoor air pollution (IAP) is lacking[16].

There are few data on adults. For rhinitis, the data in the literature indicate the existence of a relationship between exposure to visible mould and the risk of allergic rhinitis, with an increase in risk in exposed individuals that is relatively consistent between studies. However, most of these studies are based on estimates from cross-sectional surveys. A recent cohort study involving 11,506 adults from Iceland, Norway, Sweden, Denmark, and Estonia who answered a questionnaire at baseline and ten years later found that dampness and mould at home and work are associated with the onset and remission of respiratory symptoms, asthma, and rhinitis. Baseline water damage, floor dampness, mould, and mould odour were linked to the onset of respiratory symptoms and asthma, with odds ratios (ORs)[32] ranging from 1.23 to 2.24[33].

Other Physical Health Effects

There is some growing evidence that damp and/or mould may be associated with non-respiratory problems, such as throat, eye, and skin irritations and infections, nausea, fever, and tiredness[34]. Additionally, exposure to damp and mould has been linked to insomnia, snoring, and excessive daytime sleepiness[35]. Initial data suggest an association between long-term exposure to moulds and impaired cognitive function[36].

Effects on Mental Health

Dampness and mould in homes can have significant psychological effects, including stress, anxiety, and social isolation, particularly when mould is visible or ambient temperatures are cold[37]. A recent review highlights that damp and mould exposure in homes can negatively impact mental health, with studies noting links to depressive symptoms and poor self-reported mental health in adults, as well as emotional dysregulation and symptoms in children[38]. The mental health impact of mould exposure varies with the extent, location, and duration of exposure; for example mould in a bedroom, where occupants spend long periods, may pose a greater risk than in a bathroom[39]. Mould in living areas can also deter social interactions, contributing to isolation and insecurity. A study in the USA has shown an association between dampness and depression, with the effect mediated by perceptions of control over one's home and physical health[40]. A UK House of Commons Library report, citing the CDC, links living in cold or damp homes to poorer mental health, as financial stress may restrict social activities, heightening loneliness, anxiety, and depression[41,42]. The House of Commons Library paper also referred to an online survey by the mental health charity Mind, which showed that 79% of respondents with mental health issues reported worsening symptoms due to housing conditions. Among social housing tenants, the impact of damp, mould, and cold was linked to difficulty affording energy bills, further affecting well-being and health[43].

Preliminary animal studies suggest mould inhalation may impair memory and induce anxiety-like behaviour through immune responses, though further research, including biomarker and MRI studies, is needed to confirm these effects in humans.

Future studies should account for pre-existing mental health conditions to strengthen generalizability and clarify the order of causation between dampness and mental health outcomes[38].

Health Benefits of Remediation

Few studies evaluating remediation to reduce dampness and mould in buildings have been published. Assessing the effectiveness of these measures is particularly complex because of the varying initial situations and the heterogeneous nature of the renovation work. As previously discussed, the effects of remediation on the indoor air quality may not be easily evaluated using microbial and toxicological parameters[44]. A Cochrane review published in 2015 assessed the effectiveness of remediating buildings damaged by dampness or mould to prevent or reduce respiratory symptoms, infections, or asthma[45]. Only 12 randomized controlled studies were considered, most with different designs. Remediation ranged from simple cleaning to complete renovation. Pooled analysis showed a decrease in asthma-related symptoms in adults and respiratory infection following remediation. In the UK, a tailored package of housing improvements to reduce mould in the homes of children with moderate or severe asthma found 29% of those with severe asthma improved to moderate asthma within 12 months[46].

An initiative in the USA whereby structured assessments were made within homes and improvements to damp/mould, heating and ventilation were shown to result in improved quality of life, reduction in symptoms, and reduction in costs associated with hospitalization due to asthma[47]. Clinical improvement with decreased number of hospitalizations for asthma and lower consumption of anti-asthma drugs was observed in 27 patients that benefited from a home counsellor visit with allergens measures and adapted advice for global allergens avoidance. In this study, dampness markers slightly improved with an improvement of the fungal loads in two-thirds of the dwellings[48]. Moving families to "asthma-friendly" homes with moisture-proof exteriors, ventilation, and heating systems improved nocturnal symptoms in children with asthma[49]. While completely eliminating exposure to excessive mould exposure may not feasible, the remediation with the removal of the source of excessive mould is a key strategy to reduce respiratory symptoms and severe asthma[50].

Possible Actions Involving Healthcare Practitioners as Part of a Global Approach

While research has not yet determined specific thresholds for concern, addressing indoor dampness is urgent and could significantly reduce the global burden of respiratory and allergic diseases. The health effects of dampness and mould have a bearing on actions taken by local housing authorities (see *Chapter 6*). The Association of Directors of Public Health published the London Damp and Mould Checklist[51] in February 2024 is designed for use by health and social care professionals who visit residential properties as part of their management and care of patients. It provides a checklist and guidance to support the identification of internal damp and mould as well as people at risk of poor health due to damp and mould exposure in their homes.

Where concerns are identified, this resource provides guidance on actions to take in the form of advice, signposting, and template letters to inform local authority housing teams, housing associations, landlords, and health services of any concerns. This is not intended as a tool for use beyond health and care professionals (e.g. landlords) owing to its focus on assessing clinical vulnerability alongside housing concerns. However, this checklist may also serve as a useful awareness-raising tool among other frontline staff working in services beyond health and social care.

Damp and mould issues cannot be tackled in silo[52]. A life-course and holistic approach is needed, as underlined in the foundational Marmot Review[53]. Health professionals should be aware of the possibility of mould health effects and of a dedicated medical workup for diagnosis and remediation[17, 54]. Educational programmes for both health professional and patients should be considered. High-risk populations as children with asthma and/or rhinitis or immunosuppressed patients should be particularly targeted. For these populations, home surveys should be implemented taking into account the housing context (presence of visible mould, dampness, water damage, or musty odours) and the medical context (clinical signs or worsening/exacerbation of symptoms, conditions for return home in immunocompromised patients who have undergone a transplant). Current evidence does not support measuring specific indoor microbiologic factors to guide health-protective actions. Instead, addressing visible signs of dampness and mould and improving overall housing conditions are recommended[18]. The implementation of information measures aimed at the general population to help them play an active role in preventing the development of mould in the home could also be considered. Remediation cost-effectiveness could be significant as the economic cost of rhinitis and asthma is high[55,56]. The BRE has suggested that dealing with just all Category 1 hazards of damp and mould growth would save the NHS more than £33 million per year and the average cost to deal with these is less than £4,000[57].

Notes and References

1 Kearsley J. (2022) *Prevention of future deaths report: Awaab Ishak*. Available at https://www.judiciary.uk/wp-content/uploads/2022/11/Awaab-Ishak-Prevention-of-future-deaths-report-2022-0365_Published.pdf

2 Anonymous. (2022) Awaab Ishak and politics of mould in the UK. *eClinicalMedicine*. 54:101801. https://doi.org/10.1016/j.eclinm.2022.101801

3 Meadows S, Stone J, et al. (2024) The impact of the cost-of-living crisis on population health in the UK: Rapid evidence review. *BMC Public Health*. 24:561. https://doi.org/10.1186/s12889-024-17940-0

4 UK Health and Climate Change Committee. (2023) *Indoor air quality and health risks*. Available at https://assets.publishing.service.gov. uk/media/65704f719462260721c569ca/HECC-report-2023-chapter-5-indoor-air-quality.pdf

5 Vardoulakis S, Dimitroulopoulou C, et al. (2015) Impact of climate change on the domestic indoor environment and associated health risks in the UK. *Environ Int*. 85:299–313. https://doi.org/10.1016/j.envint.2015.09.010

6 Holme JA, Øya E, et al. (2020) Characterization and pro-inflammatory potential of indoor mold particles. *Indoor Air*. Jul;30(4):662–681. https://doi.org/10.1111/ina.12656. Epub 2020 Mar 18. PMID: 32078193

7 https://www.anses.fr/fr/system/files/AIR2014SA0016Ra.pdf

8 Al Hallak M, Verdier T, et al. (2023) Fungal contamination of building materials and the aerosolization of particles and toxins in indoor air and their associated risks to health: A review. *Toxins (Basel)*. Feb 25;15(3):175. https://doi.org/10.3390/toxins15030175. PMID: 36977066; PMCID: PMC10054896

9 WHO (World Health Organization) Europe. (2009) *WHO guidelines for indoor air quality: Dampness and mould*, World Health Organization, Copenhagen. Available at https://iris.who.int/bitstream/handle/10665/16 4348/9789289041683-eng.pdf?sequence=1

10 Coulburn L, Miller W. (2022) Prevalence, risk factors and impacts related to mould-affected housing: An Australian integrative review. *Int J Environ Res Public Health*. 2022 Feb 7;19(3):1854. https://doi.org/10.3390/ ijerph19031854. PMID: 35162876; PMCID: PMC8835129

11 UK Government, Department for Energy Security and Net Zero and Department for Business, Energy and Industrial Strategy. (2017) *Energy Follow Up Survey (EFUS) 2017 reports*. Available at www.gov.uk/ government/publications/energy-follow-up-survey-efus-2017-reports

12 Loukou E, Jensen NF, et al. (2024) Damp buildings: Associated fungi and how to find them. *J Fungi (Basel)*. 2024 Jan 27;10(2):108. https://doi. org/10.3390/jof10020108. PMID: 38392780; PMCID: PMC10890273

13 Lee A, Sinha I, et al. (2022) *Fuel poverty, cold homes and health inequalities*, Institute of Health Equity, London.

14 Telfar-Barnard L, Bennett J, et al. (2019) Evidence base for a housing warrant of fitness. *SAGE Open Med*. Apr 8;7:2050312119843028. https://doi.org/10.1177/2050312119843028. PMID: 31001424; PMCID: PMC6454639

15 Lorentzen JC, Johanson G, et al. (2022) Overcrowding and hazardous dwelling condition characteristics: A systematic search and scoping review of relevance for health. *Int J Environ Res Public Health*. 2022 Nov 23;19(23):15542. https://doi.org/10.3390/ijerph192315542. PMID: 36497612; PMCID: PMC9736286

16 Holden KA, Lee AR, et al. (2023) The impact of poor housing and indoor air quality on respiratory health in children. *Breathe (Sheff).* Jun;19(2):230058. https://doi.org/10.1183/20734735.0058-2023. Epub 2023 Aug 15. PMID: 37645022; PMCID: PMC10461733

17 Hurraß J, Nowak D, et al. (2024) Indoor mold. *Dtsch Arztebl Int.* Apr 19;121(8):265–271. https://doi.org/10.3238/arztebl.m2024.0018. PMID: 38381662; PMCID: PMC11381209

18 Firacative C. (2020) Invasive fungal disease in humans: Are we aware of the real impact? *Mem Inst Oswaldo Cruz.* Oct 9;115:e200430. https://doi.org/10.1590/0074-02760200430. PMID: 33053052; PMCID: PMC7546207

19 Mendell MJ, Adams RI. (2022) Does evidence support measuring spore counts to identify dampness or mold in buildings? A literature review. *J Expo Sci Environ Epidemiol.* Mar;32(2):177–187. https://doi.org/10.1038/s41370-021-00377-7. Epub 2021 Sep 2. PMID: 34475494

20 Clark SN, Lam HCY, et al. (2023) The burden of respiratory disease from formaldehyde, damp and mould in English housing. *Environments.* 10(8):136. https://doi.org/10.3390/environments10080136

21 Barnes C. (2019) *Fungi and Atopy. Clin Rev Allergy Immunol.* Dec;57(3):439–448. doi: 10.1007/s12016-019-08750-z. PMID: 31321665.

22 For more on aspergillosis See https://www.nhs.uk/conditions/aspergillosis/

23 MHCLG, DHSC & UKHSA. (2024) *Understanding-and-addressing-the-health-risks-for-rented-housing-providers/understanding-and-addressing-the-health-risks-of-damp-and-mould-in-the-home* - Updated in Aug 2024. Available at https://www.gov.uk/government/publications/damp-and-mould-understanding-and-addressing-the-health-risks-for-rented-housing-providers/understanding-and-addressing-the-health-risks-of-damp-and-mould-in-the-home–2#key-messagesMHCLG

24 Institute of Medicine (US). (2004) *Committee on damp indoor spaces and health. Damp indoor spaces and health,* National Academies Press (US), Washington, DC. Available at https://www.ncbi.nlm.nih.gov/books/NBK215643/, https://doi.org/10.17226/11011

25 Mendell MJ, Mirer AG, et al. (2011) Respiratory and allergic health effects of dampness, mold, and dampness-related agents: A review of the epidemiologic evidence. *Environ Health Perspect.* Jun;119(6):748–756. https://doi.org/10.1289/ehp.1002410. Epub 2011 Jan 26. PMID: 21269928; PMCID: PMC3114807

26 Kanchongkittiphon W, Mendell MJ, et al. (2015) Indoor environmental exposures and exacerbation of asthma: An update to the 2000 review by the Institute of Medicine. *Environ Health Perspect.* Jan;123(1):6–20. https://doi.org/10.1289/ehp.1307922. Epub 2014 Oct 10. PMID: 25303775; PMCID: PMC4286274

27 National Institute for Health and Care Excellence. (2020) *NICE guideline (NG149) Indoor air quality at home.* Available at https://www.nice.org.uk/guidance/ng149/resources/indoor-air-quality-at-home-pdf-66141788215237

28 Agache I, Canelo-Aybar C, et al. (2024) The impact of indoor pollution on asthma-related outcomes: A systematic review for the EAACI guidelines on environmental science for allergic diseases and asthma. *Allergy.*

Jul;79(7):1761–1788. https://doi.org/10.1111/all.16051. Epub 2024 Feb 17. PMID: 38366695

29 Thacher JD, Gruzieva O, et al. (2017) Mold and dampness exposure and allergic outcomes from birth to adolescence: Data from the BAMSE cohort. *Allergy.* Jun;72(6):967–974. https://doi.org/10.1111/all.13102. Epub 2016 Dec 29. PMID: 27925656; PMCID: PMC5434946

30 Varga MK, Moshammer H, Atanyazova O. (2024) Childhood asthma and mould in homes – A meta-analysis. *Wien Klin Wochenschr.* https://doi.org/10.1007/s00508-024-02396-4

31 Groot J, Nielsen ET, et al. (2023) Exposure to residential mold and dampness and the associations with respiratory tract infections and symptoms thereof in children in high income countries: A systematic review and meta-analyses of epidemiological studies. *Paediatr Respir Rev.* Dec;48:47–64. https://doi.org/10.1016/j.prrv.2023.06.0

32 The odds ratio (OR) is a measure of how strongly an event is associated with exposure. It is a ratio of two sets of odds: the odds of the event occurring in an exposed group versus the odds of the event occurring in a non-exposed group. ORs are commonly used to report case-control studies. It helps identify how likely an exposure is to lead to a specific event. The larger the odds ratio, the higher odds that the event will occur with exposure. An OR less than one imply the event has fewer odds of happening with the exposure. (See Tenny S, Hoffman MR. (2024) Odds ratio. In: *StatPearls* [Internet], StatPearls Publishing, Treasure Island (FL). Available at https://www.ncbi.nlm.nih.gov/books/NBK431098/).

33 Wang J, Pindus M, et al. (2019) Dampness, mould, onset and remission of adult respiratory symptoms, asthma and rhinitis. *Eur Resp J.* 53(5):1801921.

34 Zhang X, Norbäck D, et al. (2019) Dampness and mold in homes across China: Associations with rhinitis, ocular, throat and dermal symptoms, headache and fatigue among adults. *Indoor Air.* Jan;29(1):30–42. https://doi.org/10.1111/ina.12517. Epub 2018 Nov 2

35 Wang J, Janson C, et al. (2020) Dampness and mold at home and at work and onset of insomnia symptoms, snoring and excessive daytime sleepiness. *Environ Int.* Jun;139:105691. https://doi.org/10.1016/j.envint.2020.105691. Epub 2020 Apr 6. PMID: 32272294

36 Harding CF, Pytte CL, et al. (2020) Mold inhalation causes innate immune activation, neural, cognitive and emotional dysfunction. *Brain Behav Immun.* Jul;87:218–228. https://doi.org/10.1016/j.bbi.2019.11.006. Epub 2019 Nov 18. PMID: 31751617; PMCID: PMC7231651

37 Brooks SK, Patel S, et al. (2023) Psychological effects of mould and damp in the home: Scoping review. *Hous Stud.* 1–23. https://doi.org/10.1080/02673037.2023.2286360

38 Gatto MR, Mansour A, et al. (2024) A state-of-the-science review of the effect of damp- and mold-affected housing on mental health. *Environ Health Perspect.* Aug;132(8):86001. https://doi.org/10.1289/EHP14341. Epub 2024 Aug 20. PMID: 39162373; PMCID: PMC11334706

39 Ormandy D. (2023) Condensation and mould: Guidance for housing lawyers. *Legal Action*, Mar 2023. Available at https://lag.org.uk/article/213721/condensation-and-mould-guidance-for-housing-lawyers

40 Shenassa ED, Daskalakis C, et al. (2007) Dampness and mold in the home and depression: An examination of mold-related illness and perceived control of one's home as possible depression pathways. *Am J Public Health.* Oct;97(10):1893–1899. https://doi.org/10.2105/AJPH.2006.093773. Epub 2007 Aug 29. PMID: 17761567; PMCID: PMC1994167

41 House of Commons Library. (2023) *Health inequalities: Cold or damp homes.* Available at https://researchbriefings.files.parliament.uk/documents/CBP-9696/CBP-9696.pdf

42 US Centers for Disease Control and Prevention. (2023) *Loneliness and social isolation linked to serious health conditions.* Available at https://www.cdc.gov/aging/publications/features/lonely-older-adults.html

43 Boomsma C, Pah S, et al. (2017) "Damp in bathroom. Damp in back room. It's very depressing!" exploring the relationship between perceived housing problems, energy affordability concerns, and health and well-being in UK social housing. *Energy Policy.* 106:382–393.

44 Huttunen K, Rintala H, et al. (2008). Indoor air particles and bioaerosols before and after renovation of moisture-damaged buildings: The effect on biological activity and microbial flora. *Environ Res.* Jul;107(3):291–298. https://doi.org/10.1016/j.envres.2008.02.008. Epub 2008 May 6. PMID: 18462714

45 Sauni R, Uitti J, et al. (2011) Remediating buildings damaged by dampness and mould for preventing or reducing respiratory tract symptoms, infections and asthma. *Cochrane Database Syst Rev.* Sep 7;(9):CD007897. https://doi.org/10.1002/14651858.CD007897.pub2. Update in: *Cochrane Database Syst Rev.* 2015 Feb 25;(2):CD007897. https://doi.org/10.1002/14651858.CD007897.pub3. PMID: 21901714

46 Woodfine L, Neal RD, et al. (2011) Enhancing ventilation in homes of children with asthma: Pragmatic randomised controlled trial. *Br J Gen Pract.* Nov;61(592):e724–32. https://doi.org/10.3399/bjgp11X606636. PMID: 22054336; PMCID: PMC3207090

47 Takaro TK, Krieger J, et al. (2011) The breathe-easy home: The impact of asthma-friendly home construction on clinical outcomes and trigger exposure. *Am J Public Health.* Jan;101(1):55–62. https://doi.org/10.2105/AJPH.2010.300008. PMID: 21148715; PMCID: PMC3000722

48 Gangneux JP, Bouvrais M, et al. (2020) Asthma and indoor environment: Usefulness of a global allergen avoidance method on asthma control and exposure to molds. *Mycopathologia.* 185:367e71.

49 Kercsmar CM, Dearborn DG, et al. (2006) Reduction in asthma morbidity in children as a result of home remediation aimed at moisture sources. *Environ Health Perspect.* Oct;114(10):1574–1580. https://doi.org/10.1289/ehp.8742. PMID: 17035145; PMCID: PMC1626393

50 Denning DW, Pfavayi LT. (2023) Poorly controlled asthma - easy wins and future prospects for addressing fungal allergy. *Allergol Int.* Oct;72(4):493–506. https://doi.org/10.1016/j.alit.2023.07.003. Epub 2023 Aug 4. PMID: 37544851

51 See: https://www.adph.org.uk/resources/the-london-damp-and-mould-checklist/
52 Benton E. (2024) Damp and mould – the big picture. How do we tackle the damp and mould crisis in social housing: Lessons from the UK. *Front Environ Health*. 09 May 2024 Sec. Housing Conditions and Public Health Volume 3–2024 |https://doi.org/10.3389/fenvh.2024.1340092
53 Marmot M, Geddes I, et al. (2011) *The health impacts of cold homes and fuel poverty*, Friends of the Earth & the Marmot Review Team;201(1), London. Available at https://www.instituteofhealthequity.org/resources-reports/the-health-impacts-of-cold-homes-and-fuel-poverty/the-health-impacts-of-cold-homes-and-fuel-poverty.pdf
54 Chew GL, Horner WE, et al. (2016) Environmental allergens workgroup. Procedures to assist health care providers to determine when home assessments for potential mold exposure are warranted. *J Allergy Clin Immunol Pract*. May–Jun;4(3):417–422.e2.   https://doi.org/10.1016/j.jaip.2016.01.013. Epub 2016 Mar 25. PMID: 27021632; PMCID: PMC4972026
55 Mudarri DH. (2016) Valuing the economic costs of allergic rhinitis, acute bronchitis, and asthma from exposure to indoor dampness and mold in the US. *J Environ Public Health*. 2386596. https://doi.org/10.1155/2016/2386596. Epub 2016 May 29. PMID: 27313630; PMCID: PMC4903120
56 Rodrigo CH, Singal K, et al. (2024) Effectiveness of financial support interventions to reduce adverse health outcomes among households in fuel poverty in the United Kingdom. *Public Health Pract (Oxf)*. May 10;7:100503. https://doi.org/10.1016/j.puhip.2024.100503. PMID: 38817637; PMCID: PMC11137583
57 Garrett H, Mackay M, et al. (2023) *The cost of ignoring poor housing*, Building Research Establishment, Garston, Watford, UK.

5   Identifying and Remedying the Cause(s) (and Effects) of Dampness

Ultimately, and most importantly in order to protect health and avoid (further) damage to the structure, it is necessary to remedy the causes and sources of dampness and prevent its recurrence. The threats to the health of the occupiers cannot be removed and recurrence prevented unless the cause/source of the dampness is resolved. Once the problem of dampness is remedied, then the threats to health should be safely removed. In the case of condensation mould then it may be necessary to remove the threat to health from the mould before the work to address the underlying cause of the dampness, remembering that the issue of high RH and house dust mites might remain.

Here, we consider dampness relating to structural problems, and damp relating to energy inefficiency can be complex and is considered separately below, as is damp following flooding.

The causes of dampness result from various factors, including:

*   Inappropriate construction materials
*   Poor or inappropriate design
*   Poor or incompetent workmanship in the construction or alteration
*   A lack of proper and thorough repair and maintenance
*   Energy inefficiency (including thermal insulation, space heating, and ventilation)
*   Flooding

This chapter should be read in conjunction with Annexes 1 and 2.

Identifying the Source of the Dampness

First, there should be a general inspection of the property (on which see *Annex 1*). If the dwelling is owner-occupied this will probably be carried out by the occupier, but a full inspection and investigation should be made by an experienced professional. This is particularly so if the dwelling is of non-traditional, unusual, or unknown construction, for example large panel

DOI: 10.1201/9781003586852-5

construction, where the building is constructed of large prefabricated concrete panels or slabs. In rural areas, there are also some other forms of construction of older properties such as cob (a natural building material made from subsoil or clay, water, fibrous organic material (typically straw), and sometimes lime, which needs to contain moisture to maintain its structural strength). Unless the inspection is carried out by or on behalf of the owner this will largely be a non-intrusive inspection.

If the dwelling is rented (either from a private or a public landlord), it is in that landlord's interest to protect their property (their investment) as well as protect the tenant's health, by ensuring that any dampness is remedied.

The inspection should begin with the exterior, covering the potential weak areas as mentioned in the section on causes and structural effects, focusing on the physical structure and fabric of the property. Each element of the fabric structure (and of the dwelling as a whole) should serve its intended purpose. Any apparent design faults or evidence of poor or inadequate workmanship or maintenance should be identified and remedied.

Each individual element has a "lifespan", some much shorter than others. This means that some elements will need replacement and renewal before others. This is often taken as an opportunity to upgrade an element, but by doing this, the implications and possible effect on the dwelling should be taken into account. A dwelling is a dynamic unit, and changing one element or system may affect another. For example, the replacement of single-glazed windows with double-glazed ones may reduce the heat loss through the window areas but may interfere with the ventilation; and upgrading or changing the heating system may have similar effects. Efforts to improve air tightness can have unintended consequences including on air quality if other aspects of the dwelling are not taken into account.

For a traumatic problem, if major or sudden, it is usually relatively easy to identify the source. Comparatively more problematic may be slow leaks, where the water "runs" down internal elements (this also applies to problems of slow penetration through the roof, the water running down rafters as in Figures A1.2 and A1.3). Nonetheless, the source of the dampness must be traced and rectified.

Dampness may however be the result of flooding or new construction work. Flooding is likely to increase as a problem as the result of global heating. Over 5 million households and businesses in England – that is 1 in 6 – are currently at risk of flooding with 2.4 million properties from river or coastal flooding and 3 million from surface water, according to the BRE and Environment Agency[1]. A property can be at risk even if not close to a watercourse. Surface water flooding (pluvial) is as great a risk as river or coastal flooding (fluvial). Only once the floodwater has retreated can a dwelling be fully inspected. The dwelling can take many weeks or months to dry out fully. As we saw in *Chapter 2*, the drying out period applies to new construction too, even though less water is used these days in construction, from 2,000 to 8,000 litres can be used in building a house. It can take 12 months to two years for the house to dry out so that it is not too dry and there is a healthy RH. After flooding, care is needed throughout the drying-out process with the moisture levels monitored, especially if heater and/or dehumidifiers are used and there are hidden timbers that may not dry out as quickly as visible elements. There is a risk of a dry rot outbreak. During the winter, any moisture-saturated materials can be further damaged through frosts and the freezing of that moisture.

Dampness Related to the Design, Construction, or Repair

Where the cause of the problem is attributable to inappropriate materials, incompetent workmanship, a lack of repair and maintenance, or inappropriate design, the solution will involve replacement and may even involve reconstruction.

Determining the cause and source of dampness attributable to these matters will involve an inspection of the dwelling. This can be carried out by the occupier, but ideally an inspection should be carried out by someone with specialist knowledge. Whoever undertakes the inspection should at the least follow the process outlined in *Annex 1*.

When it comes to condensation what should be borne in mind and is worth reiterating is that merely increasing heat input and ventilation (opening windows) is not the answer, it may be economically unsound and even increase the risk of condensation.

Energy Efficiency

One of the most difficult problems influencing the likelihood of dampness is the thermal/energy efficiency of the dwelling, particularly older dwellings. The heat losses from a dwelling will be an important consideration particularly in an investigation of condensation and some information on this is included in *Annex 2*. Determining the most appropriate solution is a specialist task and this book can only provide some of the basic information.

As stated previously, it is important to note that dwellings are, or should be, designed and constructed for the purpose of occupation, and that means that they should be capable of handling the moisture produced by normal biological and domestic activities without problems occurring, such as condensation and/or mould growth. However, that capability may be compromised when alterations (including perceived "upgrading" or "modernising") are carried out. In addition, over the years, there will probably be changes to bathing/showering practices; to clothes laundering and drying; to heating, and to the type of energy used (electricity, gas, or solid fuel). These changes will all affect the amount of moisture produced within a dwelling, the ventilation, and how the dwelling copes with moisture generally. It is worth noting that the average person in England and Wales used 140 litres of water per day as of 2023. This means that a household of four could potentially use more than 550 litres of water a day[2] and it is reported that the amount of water used by the average household in the UK increased by 70% from 1985 with almost 50% used in the home[3].

Additionally, the changes in climate, such as extreme weather events, as previously mentioned, may also affect whether a dwelling is able to cope with the moisture generated without adaptations.

Many of the fundamental biological and domestic activities and functions will generate moisture. The amounts emitted will depend on the size and composition of the household, and the amount of time spent in the dwelling (on this, see *Chapter 2*). For example, a household of two elderly individuals or a household with small children would spend more time indoors than a household consisting of a working couple. As changes in occupation are likely over the years, a dwelling should be designed to be capable of being occupied by a spectrum of households.

Remedying condensation problems involves ensuring there is the right balance between thermal insulation, space heating, and ventilation. This is a specialist task.

Mould Growth

If it is not possible or practicable to remedy the main underlying structural problem immediately, and if the dwelling is occupied, it will be necessary to deal with the immediate threats to health, such as mould. All areas should be checked for mould growth, including "dead" areas (areas where there is little air movement, stagnant areas), such as surfaces hidden normally by furniture (e.g. wardrobes).

While it may be possible for very small patches of mould to be treated by the occupier using a fungicidal wash, ideally, mould removal should be carried out by a specialist. Whatever the option, there are important considerations to take into account. These include:

- The size of the infected area. If this is extensive or in several areas, the area(s) should be cleaned and treated.
- Whoever tackles the problem should wear overalls (or similar protective clothing) and a mask. If mould is disturbed, it will release spores, so as well as the protective clothing and mask, the room or area should be well-ventilated.
- Children, elderly, immuno-compromised, allergic, and asthmatic individuals should not be present within the dwelling until the cleansing is safely completed.

While not usually necessary, if it is considered needed or appropriate for the mould species to be identified, this will involve sampling. This should be carried out by a suitably trained person.

House Dust Mites (HDM)

The most common species of HDM are *Dermatophagoides pteronyssinus* and *Glycophagus domesticus*. To reduce the HDM population, the first issue is to reduce the RH, which is the fundamental cause that also led to the mould growth (and discussed throughout this book). In addition to the construction and design issues already discussed, it is advisable to keep doors to bathrooms and kitchens closed when washing and cooking. This is why it is important that the doors are in good repair and fit soundly.

Although household cleaning such as via a vacuum cleaner will not reduce HDM populations significantly, it will remove allergens, so the design and construction of the dwelling should be capable of being cleaned[4].

Notes and References

1 Environment Agency. (2017) *Building flood resilience into the fabric of Britain*. See https://environmentagency.blog.gov.uk/2017/02/17/building-flood-resilience-into-the-fabric-of-britain/andhttps://bregroup.com/about/science-park/flood-resilient-repair-house
2 https://www.statista.com/statistics/1211708/liters-per-day-per-person-water-usage-united-kingdom-uk/
3 https://www.internetgeography.net/topics/how-has-the-demand-for-water-in-the-uk-changed/
4 Crowther D, Wilkinson T. (2008) House dust mites. In: Bonnefoy X, Kampen H, Sweeney K. Eds. *Public health significance of urban pests*, WHO, Copenhagen.

6 Overview of Dampness and the Law

As a result of feedback on the first edition and other discussions, we have now included a new chapter. This new chapter is an attempt to outline various legal measures to address damp housing conditions that apply primarily in England and Wales and to debunk some misunderstandings. These measures range from contractual breaches to provisions available to local authorities to make owners improve conditions. It is intended to help those who come face to face with dampness problems but may have little understanding of the relevant legal provisions. In doing so, it also highlights the legal liabilities and responsibilities of housing providers.

One key issue is the use of the Housing Health and Safety Rating System (HHSRS), in which damp and mould growth is one of the hazards. One misconception that has been brought to the authors' attention is that damp and mould growth can only rarely be assessed as a Category 1 hazard under the Housing Act 2004. The presence of a Category 1 hazard would also mean the dwelling is non-decent (it fails to meet the Decent Homes Standard (DHS)). That said, the government has pointed out that serious Category 2 hazards of damp and mould should not be ignored.

While the evidence might not be up to date in the HHSRS Operating Guidance of 2006,[1] the Worked Examples published at the same time demonstrated Category 1 hazards for damp and mould, yet it seems that some users of the system seem to think erroneously that this could not be the case, even though we now have more evidence on the threats from damp and mould. Chapter 1 of the Operating Guidance strongly advises that users of the system keep up to date with housing and health research, so that has not happened. Unfortunately, to reiterate, those who should be in a position to provide research updates for environmental health practitioners, such as the government and professional bodies, chose not to do so.

As it stands, where the local housing authority is also the landlord, it cannot enforce notices against itself,[2] and so tenants of local authorities have to find ways of taking their own action. These routes are covered here but can also include a complaint to the Housing Ombudsman (as highlighted in the Foreword)[3] or one assumes, the Regulator of Social Housing.

DOI: 10.1201/9781003586852-6

Statutory Nuisance (Environmental Protection Act 1990)

Statutory Nuisance is a long-standing provision currently in the 1990 Act and deals with a range of matters that can constitute a nuisance or be prejudicial to health, such as noise from premises, but the most relevant so far as damp housing is concerned is:

> *any premises in such a state as to be prejudicial to health or a nuisance.*
> [s79(1)(a)]

In this context, it is the "prejudicial to health" aspect that is considered, as to be a nuisance, the condition has to affect another property.[4] It is not the disrepair or defect itself that matters but the effect or result of the disrepair or defect.[5] The relevant Statutory Nuisance provisions apply in England, Wales, and Scotland (although in Scotland, it is the Sheriff rather than magistrates' court that deals with cases). In Northern Ireland, statutory nuisance is addressed in the Clean Neighbourhoods and Environment Act (Northern Ireland) 2011.

"Prejudicial to health" is defined as meaning "injurious, or likely to cause injury, to health" [s79 (7)] but does not include "physical injury".[6] It will be satisfied where it can be shown that the state of the premises as a whole is such as would cause the health of even a sick person to deteriorate further.[7] Minor items, which alone might be relatively unimportant, can affect the overall condition of the premises so as to make them a risk to health and a statutory nuisance. Dampness has been held to make premises prejudicial to health; see, for example, *Birmingham DC v Kelly and others* (1985) 17 HLR 572.

The term "prejudicial to health" should be determined by way of an objective test, and personal circumstances of a particular individual's health should be ignored. In *Cunningham v Birmingham City Council* (1997) 30 HLR, the premises were occupied by an autistic child whose health could have been compromised by the state of the premises, but this did not fall within the meaning of the term within this legislation.

In *Birmingham CC v Oakley* [2000] 3 WLR 1936, [2001] 1 All ER 385, [2000] UKHL 59, the Justices were satisfied that premises were laid out on the ground floor in such a way that there was a risk to health, as users of the small W.C. compartment could only wash their hands in the kitchen sink or in the bathroom reached by passing through the kitchen. After an appeal was dismissed in the Divisional Court, the appeal to the House of Lords, by a majority of three to two, decided that for premises to be "in such a state as to be prejudicial to health", there had to be a feature of the dwelling that was in itself a risk to health or a source of infection or disease. Examples given were dampness, mould, and dirt or rat infestation. However unsatisfactory the design in this case, the lack of any facility or an unsatisfactory layout did not mean that dwellings were in a state that was prejudicial to health. The reasoning behind this decision was the history of public health legislation and that the premises had to have fallen or come into that state.

Where satisfied that a statutory nuisance exists, the local authority has to serve an abatement notice *R v Carrick DC, ex p Shelley* [1996] Env LR 273. Of relevance to damp and mould and the array of new and proposed provisions is that in *Shelley,* it was held that even though there were other agencies attempting to resolve the problem of sewage on a beach, the duty to serve an abatement notice still applied, and the local authority had failed in its statutory duty by resolving not to take any action.

To be of any use for tenants, the act or default must obviously be that of the landlord or owner. This provision can be used where condensation and associated mould growth occurs as the result of inadequate construction and design and poor insulation of properties of a dwelling with inefficient or ineffective heating and ventilation systems (as discussed in earlier chapters). As the burden of proof is to the criminal level, it is not likely to be the fault of the owner or landlord if the cause of condensation and dampness is the non-use of heating appliances by the occupier (which is why the investigation has to be thorough), and the heating system is adequate for the property and not defective or so expensive such that no reasonable "landlord" could expect any occupier to use it [*Dover District Council v Farrar* (1980) 2 HLR 32; *GLC v London Borough of Tower Hamlets* (1983) 15 HLR 54; *Birmingham DC v Kelly* (1985) 17 HLR 572]. The abatement notice must specify the steps necessary to remove the nuisance and a reasonable time for them to be completed. The authority may prosecute for failure without reasonable excuse to comply with the abatement notice and/or carry out the work in default and recover the costs.

Where there is a statutory nuisance, someone directly affected by the conditions may complain to the local Magistrates' Court [s 82 EPA]. Section 82(6) and (7) of the EPA requires at least 21 days' notice to be given in writing of the intention to complain (seek the issue of a summons) to the magistrates' court and to institute proceedings. The defendant to such proceedings, in the case of the nuisance arising from any defect of a structural character, is the owner of the premises (section 82(4)(b)).

When it comes to who can provide expert evidence on when premises are prejudicial to health, *Patel v Mehtab* [1980] 5 HLR 78, QBD and *O'Toole v Knowsley Metropolitan Borough Council* [1999] Env LR D29; Times Law Reports 21 May, QBD, underscored that a qualified environmental health officer could provide such evidence.

Leeds v Islington LBC (1999) 31 HLR 545 made it clear that a s.82(6) notice must be served at the "proper office" of the prospective defendant, which in the case of a corporate body like a local authority means its registered or principal office (s.160 of EPA 1990). The local housing office (even if that is the address on the tenant's handbook for complaint) is not adequate.

The works to abate that statutory nuisance can include works of improvement. See, for example, *Dingwall v Lambeth LBC* (Camberwell Green Magistrates' Court, 28 April 2014) (*Legal Action* December 2014), where the tenant's flat had mould and damp that the local authority had failed to eradicate, and the landlord was required to carry out works to improve her home so the damp and mould would not return. This work included the installation of double glazing and upgrading the insulation. In other words, the courts can require any work seen as necessary to abate the statutory nuisance, and this is not limited to works of repair.

Landlord and Tenant Act 1985 (as Amended by the Homes (Fitness for Human Habitation) Act 2018, etc.)

The provisions discussed here imply terms into tenancies, and as such are enforced by way of an action for breach of contract, which can be a claim for damages and/or a court order for specific performance.

Fitness for habitation is addressed in s.9A of the 1985 Act but is defined in s.10, which sets out the factors to be considered in determining whether a house or dwelling is unfit for human habitation and to which regard shall be had to its condition in respect of the following matters:

- Repair
- Stability
- Freedom from damp
- Internal arrangement
- Natural lighting
- Ventilation
- Water supply
- Drainage and sanitary conveniences
- Facilities for preparation and cooking of food and for the disposal of wastewater; and in relation to a dwelling in England, any hazard as prescribed by regulations made under s.2 of the 2004 Act, i.e. the HHSRS hazards which include damp and mould growth. The dwelling shall be regarded as unfit for human habitation if, and only if, it is so far defective in one or more of those matters that it is not reasonably suitable for occupation in that condition.

As s.9A was introduced by The Homes (Fitness for Human Habitation) Act, which only applies to tenants in England, this provision in the 1985 Act only applies in England. It applies to leases of seven years or less and implies a covenant by the lessor of the dwelling that it is fit for human habitation at the time the lease is granted or otherwise created, at the beginning of the term of the lease, and will remain fit for human habitation during the term of the lease.

It does not cover people who have "licences to occupy",[8] instead of a tenancy agreement. The Environmental Protection Act makes no such distinction.

A new Section 10A was introduced by the Social Housing (Regulation) Act 2023 (see "Awaab's Law" below).

The Landlord and Tenant Act also includes a repairing obligation on landlords in section 11 which implies into any tenancy of less than 7 years [see ss.13 and 14 LTA 1985] a contractual obligation on the landlord to the following:

a) Keep the structure and exterior of the dwelling house (including drains, gutters, and external pipes) in repair
b) Keep in repair and proper working order the installations in the dwelling house for the supply of water, gas, electricity, and sanitation (including basins, sinks, baths, and sanitary conveniences but not other fixtures, fittings, and appliances for making use of the supply of water gas or electricity)
c) Keep in repair and proper working order the installations in the dwelling house for space heating and heating water.

This is not available where the problem is condensation as the result of design failures or inadequate facilities if they are not in disrepair (but s.9A would remain relevant). The covenant would only apply where the condensation led to disrepair – for example, rotted window frames.[9] It is also relevant where the dampness is the result of disrepair – for example, penetrating damp as the result of leaking gutters.

For the landlord to be in breach in general, they have to have had notice (been informed) of the problem and failed to carry out the remedial work in a reasonable time. How they are notified is not so important; it could be as the result of an advice agency telling them or as the result of the service of an Awareness Notice[10] (what was a Hazard Awareness Notice; see later). Actions for breach of the repairing obligation will normally be for damages, although an order for specific performance can be made under s.17.

Action for a breach of section 9A again will normally be for damages as the result of the breach of the obligation for the dwelling to be fit (although again an order for specific performance to get work undertaken can also be sought). It is worth noting that a decision in the Central London County Court, while not binding, found that whether a dwelling was fit or unfit under s9A of the 1985 Act was, in effect, a binary decision, and, as the tenant had been deprived of any value from the tenancy during the period it was unfit, no rent was payable (*Dezitter v Hammersmith and Fulham Homes* (Central London County Court, 7 November 2023) Reported in Nearly Legal https://nearlylegal.co.uk of 23/11/23). This was also followed in *Mason v 1) Olivera and 2) Santana*, Claim no: K2PP0132 (15th December 2023, County Court at Clerkenwell and Shoreditch), where the action on conditions was a counterclaim for possession and where there had been leaks and mould growth, the local authority had also served an Emergency Prohibition Order. In *E v The London Borough of Lambeth* (Wandsworth County Court, 17 April 2024), a slightly different approach was taken. Since the start of the tenancy in 2018, there had been damp and mould due to defects to external brickwork and a failed damp proof course, but the award for the conditions did not amount to 100% of the rent, although a finding of unfitness for human habitation "should mean damages of a considerable proportion of the rent".

It is perhaps relevant that in *Bole v Huntsbuild Ltd (2009) EWCA Civ 1146* and *Rendlesham Estates v Barr (2014) EWHC 3968 (TCC),* both section 1 Defective Premises Act 1972[11] cases (see Nearly Legal 13/1/2019 https://nearlylegal.co.uk/2019/01/build-defects-and-fitness-for-habitation/), decisions were taken that have been of assistance in assessing fitness for human habitation under section 9A. In *Jillians v Red Kite Community Housing.* Oxford County Court 24 September 2024 (unreported) (see Nearly Legal, https://nearlylegal.co.uk/2024/09/the-meaning-of-unfitness/) "fitness for human habitation" was considered with assistance from those cases, in particular. Also, an expert's finding as to the hazard to health is not required for the Court to make a finding on unfitness. It was suggested that to say that a Court cannot make a finding of unfitness without expert evidence to that effect is to "mistake the role of the expert, and of the court".

Part 4 of the Renting Homes (Wales) Act 2016 sets out the obligations placed on a landlord with regard to the condition of a dwelling. Section 91 of the Act places an obligation on a landlord to ensure that, at the start of and during the length of the occupation contract, the dwelling is fit for human habitation. These obligations are set out in The Renting Homes (Fitness for Human Habitation) (Wales) Regulations 2022, which in the schedule set out the 29 matters and circumstances to which regard must be had when determining whether a property is for human habitation – essentially the HHSRS hazards.

Housing Act 2004 and Dampness

Part 1 of the 2004 Act introduced the notion of Category 1 and Category 2 hazards. The prescribed method of determining whether a Category 1 or 2 hazard exists is the Housing Health and Safety Rating System (HHSRS). The HHSRS Regulations are made under s.2[12] and describe the hazard as "Exposure to house dust mites, damp, mould or fungal growths". A Category 1 hazard is one that is assessed as having a hazard rating of 1,000 or more when using the prescribed method of assessing the threat to health and/or safety.

There is a duty on local housing authorities (LHAs) to take one of the courses of action listed below where there is a Category 1 hazard. For Category 2 hazards (those scoring 999 or less), there is a power or discretion to act, and it can be said that in theory at least showing how that discretion has been used is more onerous than when there is a clear duty in the case of a Category 1 hazard.

Section 3[13] requires the LHA to keep the housing conditions in their area under review with a view to identifying any action that may need to be taken by them under Part 1 of the Act. Section 4 supplements that so if the LHA becomes aware that it would be appropriate to inspect any "residential premises" in their district to determine whether any Category 1 or 2 hazard exists on those premises, the authority must do so in accordance with the HHSRS Regulations. A magistrate can make an "official complaint" to the LHA about the existence of a hazard and require such an inspection by the "proper officer" and preparation of a report.

The powers available under Part 1 of the Act are to serve:

* Improvement Notices (which may also be suspended)
* Emergency Remedial Action
* Prohibition Order (which may also be suspended)
* Emergency Prohibition Order
* Awareness (previously Hazard Awareness Notice)
* Demolition Order
* Clearance Area

LHAs have powers to take emergency remedial action or make emergency prohibition orders (ss. 40 and 43) where there is a Category 1 hazard, and there is an imminent risk of serious harm to health or safety of any of the occupiers, and there is not a management order in force.

Where the Improvement Notice or Prohibition Order is suspended, the suspensions must be reviewed at least annually. Improvement Notices and Prohibition Orders are registered as local land charges, but not Awareness Notices, as they merely advise the person served as to the existence of a Category 1 or Category 2 hazard or failure to meet type 1 or type 2 requirements of the DHS in qualifying residential premises (see below). These do provide evidence that the landlord has had notice of the disrepair or conditions. The hazard rating does not dictate the course of action that the local housing authority might take, only that they must act in the case of a Category 1 hazard.

Determining a hazard rating depends on the evidence gathered on an inspection of the whole dwelling and the reasoning or justification as much as actual numbers. It should be recognized that a Category 1 hazard by itself will mean the dwelling fails to meet the DHS. Secondly, as the letter from Michael Gove quoted in *Chapter 1* made clear, a hazard of damp and mould growth, even if not rated as Category 1, can still pose an unacceptable threat to the health of the occupiers, and so action should be taken.

As it has been thought by some that the tables in the Operating Guidance suggest that a Category 1 hazard for damp and mould growth is unlikely.

The reader is asked to consider the case study below (this is not intended as a full inspection report), noting that the inspection identified deficiencies that contribute to the hazard of damp and mould and that there is a high likelihood such that a Category 1 hazard exists, which could be justified, but also why it was advised that hazards in Bands D and E should not be ignored.

Case Study

A ground-floor two-bedroom flat in a purpose-built three-storey block constructed in the 1920s. There is a living room and kitchen and a combined WC/bathroom. The floor is of solid concrete construction, and the walls are 225 mm solid brick construction. The windows are uPVC double-glazed units, but not all have trickle vents. Heating is via a combination gas boiler and water-filled radiators. The room thermostat was noted as set at 21°C. The flat is occupied by two adults and two children aged 3 and 4. The orientation of the flat is East/West, with the end external wall facing north. The inspection was undertaken in early November.

The following deficiencies were reported:

1. Damp affected plaster to a low level to the external (north facing) living room wall (this is behind where the sofa is positioned and there is a substantial amount of furniture in the room)
2. Small area of mould-affected wall plaster in the first bedroom
3. Area of mould affected wall and ceiling plaster and window frame of the second (smaller) bedroom (See Figure 6.1)
4. Trickle vents, where present, blocked with dust and dirt
5. Ill-fitting bathroom door
6. Some mould on bathroom window frame (ventilation via high-level opening light to window only)
7. Fixed glazing to kitchen and extract fan with pull cord switch but cord difficult to reach over kitchen sink unit (extract fan is operational)

Rating

Factors Affecting the Likelihood of an Occurrence

Occurrence is defined as an event or period of time exposing an individual to a hazard, which itself is defined as risk of harm to the health or safety of

Figure 6.1 Mould in the smaller bedroom

an actual or potential occupier that arises from a deficiency. The likelihood is of an occurrence over the next 12 months.

The vulnerable age group is children aged 14 and under and likelihood of an occurrence that could cause harm over the next 12 months is assessed as 1 in 1 because there is mould in the bedroom that would be occupied by children of the vulnerable age group (the fact that it is actually occupied by the vulnerable age group does not figure in the assessment but could influence the course of action) and where they would spend considerable time. There is little opportunity to avoid exposure to mould spores. The evidence from the inspection is that there is high RH, which also increases the house dust mite population. The presence of visible mould will also lead to stress and mental ill-health justifying attention.

The indications are that the average likelihood of an occurrence is far greater than in the current Operating Guidance (1 in 400). In the stock as a whole of 24,408 million households, 1,009,000 have any damp (EHS 2022/23) indicates a chance of around 1 in 24 of a dwelling having any dampness (not necessarily causing ill-health justifying medical attention).

Some 455,000 dwellings had mould (the figure for 2019). There is a 1 in 54 chance of having damp and mould growth. Figures from NICE show allergic rhinitis affects 23–30% of the population in Europe, suggesting something like 14 million people could be affected by allergic rhinitis (10% of 6- and 7-year-olds and 15–19% of 13- and 14-year-olds are affected by allergic rhinitis (https://cks.nice.org.uk/topics/allergic-rhinitis/background-information/prevalence/)) in England. If only 2% of those affected were due to damp and mould, that would still be 288,600 people and (taking the number of dwellings with damp and mould growth as 455,000) indicates a chance of this allergic response as 1 in 1.5. Clark *et al.* found exposure to damp and/or mould was associated with approximately 5,000 new cases of asthma and approximately 8,500 lower respiratory infections among children and adults in 2019. Those figures indicate that as a starting point, living with damp and mould could mean at least a 1 in 34 chance of ill-health.

Allergens associated with house dust mites are the most common triggers of asthma and are also implicated as a causal agent of the illness. mould is associated with the development of asthma in young children, evidence suggesting a causal relationship. WHO reported that about 15% of new childhood asthma in Europe can be attributed to indoor dampness.

The average spread of harm outcomes is Class I 0; Class II 1; Class III 10; Class IV 89. but despite the high likelihood of an occurrence justifying medical attention, the spread of harms is largely unaltered, although evidence on health, for example on rhinitis, justifies increasing Class III harms. It might seem strange not to increase Class I harms given that there is evidence that there can be the most severe outcome, but as this is very rare by comparison with the number of damp homes it is still far less than 0.1% and for this, the Representative Scale Point is 0.

Likelihood of 1 in 1.
Outcomes: Class I – 0, Class II – 1, Class III – 21.5, Class IV – 77.5
Hazard Score: 8225. Band A

Should the assessment be that there is only a 50:50 chance of the vulnerable age group (14 years and less) being affected in this property (Likelihood of 1 in 2), the hazard score of 4113 would still be Category 1 (Band B).

Alternate Scenario Hazard Rating

Consider further, if the only mould and dampness was in the larger bedroom, arguably the likelihood would be less at 1 in 10 (10 times less likely than the above assessment but 40 times more likely than the average of 1 in 400 given in the 2006 Operating Guidance) because there is still evidence of high RH and associated threats, and with the average outcomes would give a rating of 489 which is Band E (and close to band D).

If the only damp and mould detected at the time of inspection was that in the living room (Item 1) the likelihood of an occurrence given the other deficiencies would still be greater than average and an assessment of 1 in 56 could be a conservative assessment given the time of year of inspection (near the start of the "condensation season") that would give a hazard rating of 87 (Band G).

In the previous scenario (and variations), there is no comment as to the course of action to be taken under Part 1 of the 2004 Act, which could also be strongly influenced by such things as the vulnerability of the actual occupiers both to the hazard itself and more generally and should take account of the enforcement and other relevant guidance.[14] It should be noted that if this property was owned by a registered housing provider with a Category 1 hazard, then as the result of the Renters' Rights Bill as introduced, it would be non-decent and outside the scope of the Housing Act 2004 Part 1 so far as the DHS is concerned and it would be subject to regulation by the Regulator of Social Housing.

Changes to the Law: Awaab's Law – Social Housing (Regulation) Act 2023 s.42 and Renter's Rights Bill

The Social Housing (Regulation) Act 2023 Act amended s.10 of the Landlord and Tenant Act 1985 (as previously mentioned) by introducing section 10A to do with remedying of hazards occurring in dwellings let on relevant social housing leases. The Secretary of State is required to set out new requirements for landlords to address hazards such as damp and mould within a fixed period – that is to set time limits for dealing with the identified problem. Consultation has taken place on these requirements.[15] Once this comes into force, a tenant could enforce their rights by a claim for breach of covenant (in the Landlord and Tenant Act 1985) if their landlord does not comply. From October 2025, social landlords will be forced to investigate and fix dangerous damp and mould in set time periods, as well as repair all emergency hazards within 24 hours. At the time of writing the time periods have not been set.

There is, as discussed already, a provision for action by tenants where the dwelling is considered unfit for human habitation under the amended Landlord and Tenant Act 1985.

Strangely, the consultation on Awaab's Law made no mention of, or reference to, the statutory duties and powers placed on local housing authorities (LHAs) by Part 1 of the Housing Act 2004 to deal with conditions that constitute HHSRS Hazards. This legislation is of particular significance as the duties and powers placed on LHAs have applied to all housing tenures including social housing (with the single exception of housing owned by the enforcing authority[2]).

Also of significance is that the assessment and enforcement processes available to the local housing authority are independent of the landlord, unlike the proposals made to meet "Awaab's Law". The government has suggested that this new provision will enable resident to "hold their landlords to account by taking legal action through the courts for a breach of contract." However, this should be qualified and recognize that for a tenant to challenge the landlord (with its available resources and access to experts) is a daunting prospect, that ideally the tenant should have access to specialist legal advice and independent expert support, and that a court is an intimidating arena. To obtain this advice and support is likely to be costly, and legal aid or legal advice may not be readily available (see below).

The lack of reference in the consultation, to the duties and powers placed on local authorities by Part 1 of the 2004 Act was surprising, particularly as paragraph 55 stated that ". . . Awaab's Law should take into account the 29 health and safety hazards [. . .] set out by the Housing Health and Safety Rating System" – the hazards used as the basis for triggering action under Part 1 of the 2004 Act.

There is also the issue of whether there is, or could be, a conflict between the powers and responsibilities of the Housing Ombudsman or the PRS Housing Ombudsman as proposed in the Renters' Rights Bill (and Regulator of Social Housing and on this) and the duties and powers of the local authority? What if there is a disagreement about the remedial action? Which body takes priority – the Housing Ombudsman (a national body) or the local housing authority (a local body that is responsible for dealing with housing conditions)?

It is proposed that the scope of Awaab's Law is restricted to "significant risk [. . .] to the actual resident of the dwelling". This fails to recognize that a resident may become vulnerable, that a friend or relative who is (or could be) vulnerable may visit, or that there may be a change of resident. Ideally, a dwelling should be safe for any potential or actual resident and any visitor. Then it is proposed that to ". . . determine whether a hazard poses a significant risk [. . .] landlords use their judgement and the existing processes they have in place . . .". This implies that landlords either have, or have access to, personnel trained and experienced in applying the HHSRS to assess housing conditions. One thing that has been noticeable over the years is that many social landlords or their officers are unaware of the HHSRS and may not have received any training on it.

The Renters' Rights Bill introduced into Parliament in September 2024 intends to extend Awaab's Law to the private rented sector (and also apply the DHS to the private rented sector). The difficulties in using the provisions in the amended Landlord and Tenant Act discussed earlier will remain. It could be confusing, and the time limits set in any regulations on dealing with hazards may not be suitable for all circumstances or could be the lowest common denominator. What if in a particular case the enforcing authority considers there is an imminent risk of serious harm and remedial action should be taken sooner than set out in regulations under the Landlord and Tenant Act and so uses the power to take emergency action?

Applying the DHS to the private rented sector via an amended Part 1 of the Housing Act 2004 also raises some interesting issues. The Bill gives the Secretary of State the power to make regulations specifying DHS requirements that must be met by qualifying residential premises. Qualifying residential premises are defined in the Bill and exclude social housing let by landlords who are registered providers of social housing, that is registered with the Regulator of Social Housing.

The Bill sets out a non-exhaustive list of matters that these requirements may include and. allows the requirements to be split into two categories: Type 1 requirements that a local housing authority has a duty to enforce; and Type 2 requirements that a local housing authority has a power to enforce. At the time of writing the types of requirements have not been defined. It also allows the regulations made to include exceptions from the requirements. The Bill will bring the majority of the private rented sector into scope of the DHS by covering dwellings and HMOs let under assured or regulated tenancies, and supported housing let under a tenancy or occupied under licence. What is confusing in the published Bill is that if premises provided by a social landlord fail the DHS because of the presence of a Category 1 hazard then the local housing authority cannot help the tenants by serving a notice. The Explanatory Notes published with the Bill suggest that compliance will be overseen by the Social Housing Regulator.[16] This begs the question of whether Social Housing Regulator has sufficient competent staff to deal with such cases properly. The same question arises perhaps even more when it comes to the Ombudsman for the PRS when applying "Awaab's Law" to that sector.

Legal Aid (Public Funding) and Other Considerations

Where tenants seek to secure their own legal remedies, such as a claim under the Landlord and Tenant Act, the problem can be funding of such action. According to the National Audit Office in 2024 the Ministry of Justice "has succeeded in its objective of significantly reducing [total] spending on legal aid, which has fallen by more than a quarter in the last decade in real terms".

It has been increasingly difficult for tenants to take their own action. In part, this is the result of cuts to public funding and fees that the Legal Aid Agency (LAA) will allow, and these are governed by Regulations that have not been updated for a number of years[17] (legal aid fees have not increased since 1996, and in 2011 were cut by 10%). According to the Law Society, some 42% of people in England and Wales do not have access to a local housing legal aid provider and the number of civil legal aid providers in England and Wales has reduced by 19% in the last five years.[18] Interestingly, a report from The Law Society and Frontier Economics[19] gave a specific example of why public funding should be increased, saying that 48% of current UK tenants were living with one or more housing disrepair issues, resulting in NHS treatment costs. If reforms to housing legal aid could reduce the prevalence of Category 1 hazards in rented housing by just 5%, there would be a saving to the NHS of over £15 million a year. Although the legal provision exists can be difficult for tenants to take their own action, which makes it more important for local housing authorities and regulators to use their powers effectively.

Civil legal aid, for example to make a claim for a breach of the implied terms on fitness or disrepair in a tenancy, pays for legal advice for people who cannot afford to pay for a solicitor to represent them. If an application for legal aid is successful, it can pay for some or all of their legal costs. Public funding is available for counterclaims, including disrepair counterclaims.

That said, there is no legal aid for cases taken under the Environmental Protection Act 1990 Part 3 (statutory nuisance). Legal Help can be used to investigate whether action under the Environmental Protection Act is a valid course of action, but funding is not available to take the case to Court. Alternative funding can be used to pursue the private prosecution from that point, including proceeding under a Conditional Fee Agreement (CFA). There is no damages claim in such cases (although compensation may be awarded if the landlord is found guilty).

In addition, the Civil Procedure Rules[20] can leave tenants concerned about retaliatory evictions or loss of tenancy even if s.21 is repealed. These set out procedures to be followed and could be seen as a desire to keep claims out of Court.

For civil claims, legal aid would be available in relation to the removal or reduction of a serious risk of harm to the health or safety of an individual (or a relevant member of that person's family) where:

- the risk arises from a deficiency in a home;
- the home is rented or leased from another person; and
- the services are provided with a view to securing that the other person makes arrangements to remove or reduce the risk.

"Deficiency" is defined in The Lord Chancellor's guidance under section 4 of Legal Aid, Sentencing and Punishment of Offenders Act 2012 as "any deficiency, whether arising as a result of the construction of a building, an absence of maintenance or repair, or otherwise". This definition, as well as covering deficiencies in the home arising from lack of maintenance and lack of repair will also, for example include poor design of a home that leads to a risk to health, such as condensation dampness.

The definition of "harm" in the 2012 Act includes temporary harm and the definition of "health" includes mental health, the same as in the Housing Act 2004. The Lord Chancellor's Guidance says that in some cases, for example, those involving a deficiency such as a leaky gas boiler or dangerous electrical wiring, the seriousness of the risk to the health of the client may appear a relatively clear-cut issue. In other cases, the seriousness of the risk of harm that the deficiency poses to health or safety will vary depending on the individual circumstances of the case. For example, the risk of harm to the health of a tenant who has a respiratory illness from damp may be greater than the risk to a tenant who does not.[21]

The guidance does not include a non-exhaustive list of the factors to be taken into account when deciding whether the "serious risk" requirement is met. In order to determine whether the serious risk requirement is met, the Director of the LAA will need to take into account all relevant factors. By way of guidance, included in these factors are:

- Whether the applicant or relevant family members affected by the deficiency are in a high-risk age group, such as the elderly and very young children, and therefore more susceptible to any deficiency.
- Whether the applicant is vulnerable due to a disability. For example, a leaking roof which causes flooring to be damp may be viewed as significantly more serious if the applicant has particular mobility problems.
- Whether the Local Authority has already identified hazards that arise from deficiencies in the home. For example, under the Housing Health and Safety Rating System.

There are no specific evidence requirements to be satisfied in relation to those criteria. Legal aid is available, so long as there is a credible allegation of serious risk, to enable providers to obtain the necessary evidence, including expert reports.[21]

Legal aid funding will cease to be available once the risk of harm has been removed or reduced rather than when the repairs are completed, or the case is at an end.

Legal advisors will need to be able to explain to the LAA why the disrepair case was unsuitable for funding under a CFA.[22] In practice, legal aid will be available for cases meriting interim injunction applications to deal with urgent disrepair, which by their nature are likely to satisfy the above criteria, but where any damages claim is likely to be minimal.

To actually bring a claim for a failure under the Landlord and Tenant Act (as amended) using representation paid for under legal aid, it will be necessary to establish that there is a serious risk of harm to the tenant or a member of their household for which there will need to be evidence such as could be provided by an EHP whether employed by the local authority or otherwise. While the local housing authority can serve a notice to ensure that remedial work is carried out, they cannot secure compensation for the time when the household has been exposed to the threats to health, and the 1985 Act is a route to secure compensation if that is appropriate, and the notice served by the local housing authority can be used to support a claim.[23]

Notes and references

1 Office of the Deputy Prime Minister. (2006) *Housing health and safety rating system (HHSRS) operating guidance: Housing Act 2004. Guidance about inspections and assessment of hazards given under s.9.* Available at https://assets.publishing.service.gov.uk/media/5a78d3d940f0b62b22c bd1d6/142631.pdf (This Department is now Ministry of Housing Communities and Local Government).

2 See *R v Cardiff City Council ex p Cross* [1982] 6 HLR 6.

3 See https://www.housing-ombudsman.org.uk

4 *National Coal Board v Thorne* [1976] 1 WLR 543.

5 For more detail on this see Battersby S, Pointing J. (2019) *Statutory nuisance and residential property: Environmental health problems in housing.* Abingdon: Routledge. https://www.routledge.com/Statutory-Nuisance-and-Residential-Property-Environmental-Health-Problems-in-Housing/Battersby-Pointing/p/book/9781138338135

6 *Everett, R (on the application of) v Bristol City Council* [1999] EWCA Civ 869.

7 *Malton Urban Sanitary Authority v Malton Farmers Manure Co.* [1879] 4 Ex D 302).

8 Be aware there are "sham licences" where in fact there is a tenancy – see *Street v Mountford* [1985] 17 HLR 402, HL.

9 See *Quick v Taff Ely Borough Council* [1985] [1986] QB 809, [1985] 3 All ER 321, [1985] EWCA Civ 1, 18 HLR 66, [1985] 3 WLR 981, 276 EG 452, [1985] EGLR 50, 84 LGR 498.

10 Hazard Awareness Notices under the Housing Act 2004 will become Hazard Notices as the result of amendments in the Renters' Rights Bill.

11 s.1 Defective Premises Act 1972 (Duty to build dwellings properly) – a person taking on work for or in connection with the provision of a dwelling (whether the dwelling is provided by the erection or by the conversion or enlargement of a building) owes a duty such that a the dwelling will be fit for habitation when the works is completed.

12 SI 2005 No 3208 The Housing Health and Safety Rating System (England) Regulations 2005.

13 Note the Renters' Rights Bill introduced in 2024 seeks to amend Part1 of the Housing Act 2004 by introducing provisions so that the Decent Homes Standard (DHS) will be enforced under Part 1.

14 See HHSRS Enforcement Guidance. Available at https://assets.publishing.
    service.gov.uk/media/5a7960d9e5274a3864fd6822/safetyratingsystem.pdf
15 See https://www.gov.uk/government/consultations/awaabs-law-consultation-
    on-timescales-for-repairs-in-the-social-rented-sector/b7173b41-
    1d97-495d-857a-3085f95d26ff
16 see Para 641 of Explanatory Notes. Available at https://publications.par-
    liament.uk/pa/bills/cbill/59-01/0008/en/240008en.pdf
17 See The Civil Legal Aid (Remuneration) Regulations 2013 SI 2013
    No 422.
18 https://www.lawsociety.org.uk/campaigns/civil-justice/legal-aid-deserts/
19 The Law Society & Frontier Economics. (2024) *Research on the sustain-
    ability of civil legal aid, final report*, Commissioned by the Law Society.
    Available at https://www.lawsociety.org.uk/topics/research/civil-legal-
    aid-sustainability
20 See for England https://www.justice.gov.uk/courts/procedure-rules/civil/
    protocol/prot_hou, this cover claims based on the new section 9A in Landlord
    and Tenant Act 1985 (implied term as to fitness for human habitation) which
    applies only in England and for Wales https://www.justice.gov.uk/courts/
    procedure-rules/civil/protocol/pre-action-protocol-for-housing-disrepair-
    cases-wales. These do not cover claims brought under section 82 of the
    Environmental Protection Act 1990 (which are heard in the Magistrates'
    Court).
21 *The Lord Chancellor's guidance under section 4 of Legal Aid, Sentenc-
    ing and Punishment of Offenders Act 2012* paras 12.6–12.10. Available at
    https://assets.publishing.service.gov.uk/media/64008603d3bf7f25fcdb849f/
    Update_Lord_Chancellor_s_s4_Guidance_Aug21-_FINAL.pdf
22 (Civil Legal Aid (Merits Criteria) Regulations 2013 SI No 104 reg 39(b)).
23 For more details on claims see Luba J, O'Donnell C, Peaker G. (2019)
    Housing conditions: Tenants' rights, 6th Ed, LAG, London.

Annex 1

Investigating the Causes/Sources of Dampness

This Annex provides a brief outline of the approach to investigating dampness, and it is followed by a checklist to help with that investigation.

Wall structure can influence the presence of dampness, for example a solid wall of older construction might be more susceptible to penetrating dampness than a cavity construction, particularly if the pointing is open and perished.

Signs of potential sources of penetration of dampness through the roof(s) include slipped or missing slates and cracked tiles, particularly if there is no roofing felt under the roof covering. A full examination of roofs including flashings and valleys will not be possible without ladders, although binoculars can be helpful. Similarly, the condition of flat roofs may not be possible without means of access (or a view from an adjacent higher building).

Dampness from these sources in particular will reflect the weather – the damp internally will be affected by rainfall and will be less in drier periods, although internal decoration may be stained.

Blocked, cracked, or leaking rainwater goods will also be apparent by the staining to adjacent wall surfaces (this can include salts leached out of brickwork by rainwater), but during dry periods, the wall may not be damp although a stain might persist. Water from defective eaves gutters and down pipes will erode render and brickwork (particularly the mortar joints to a brick wall). Again, the weather will have some effect on the internal dampness.

Rotting timber to window and door frames or thresholds may be a source of internal dampness, as may poor or inadequate sealing. It may also be that pointing (or mastic) may be absent from under external sills that can allow water to penetrate the structure.

Internally, the dampness or staining may be patchy, but usually (though not always) coincides with the external defect.

Where soil, paving, or a hard surface covers the damp proof course (DPC), this will negate the DPC and allow rising dampness. A DPC should be at least 150 mm (6 inches) above the adjacent soil or paving level to avoid rain bouncing above the DPC. Internally, rising dampness will create a "tide" mark to the plaster to a height of about 1 m (or around 3 feet). Mould on rising damp internally is rare as the moisture will normally contain salts drawn up from the soil, which will inhibit mould growth. However, it is not unknown for condensation to run down the surface of a wall onto skirtings and then rise up in the wall plaster causing a stain giving the superficial appearance of rising damp. It is also possible that where a problem of rising damp has been dealt with by the insertion of a DPC (perhaps in an older property that had never had one), the wall still appears damp as the result of migration of moisture to the surface and carrying salts that attract moisture from the air (hygroscopic). This will occur if the wall has not been allowed to dry out prior to replastering or a specialist salt and moisture-resistant plaster has not been used.

Damp patches can occur on chimney breasts where this is or more likely was an open fireplace, and sometimes where there is a gas fire or a solid fuel stove and an unlined flue. These result from hygroscopic salt deposits (salts that absorb moisture from the air), and where the flue has not been lined. This occurs when the products of combustion within the flue cool and the resulting condensate contain the salts which are deposited within the unlined walls of the flue. As they build up, they begin to absorb more moisture, damaging the plaster and sometimes leaving a brownish stain. Lining the flue with a waterproof material or liner prevents this problem.

The presence of mould growth with no obvious external source of moisture, usually indicates that the problem is condensation (and high RH). This affects areas that a colder than others. For example, areas above the window and door openings where a dense material is slower to warm up than adjacent areas; where there is a link across the cavity, either by mortar on wall ties or lintels above openings;[1] or at corners of walls where heat loss is higher than elsewhere.[2] Mould growth is more likely in cooler and perhaps unheated rooms and areas. It is also possible where there is a lack of air movement such as behind furniture.

Lack of air movement can also cause condensation in a roof space if the insulation is laid between the ceiling joists and there is no air movement (ventilation) above the insulation. It has been known for condensation to form on the underside of the roof covering and drip onto the insulation. This can reduce the effectiveness of the insulation and can lead to damp patches, that might be assumed to indicate a leaking roof. Access to the roof space would be necessary if this is suspected.

A visual and superficial inspection may not give definite indications as to the cause but may help to eliminate those that are less likely. A detailed inspection by a specialist may be necessary, and, in some cases, may involve opening up the structure.

An Environmental Health Officer (EHO) or EHP can only undertake a non-intrusive inspection, but the owner (or someone instructed by the owner) can open up the structure or remove fixtures or fittings.

Useful Equipment

While at an initial inspection or visit it might not be necessary to use equipment, or it might not be available, there are two pieces of equipment that an EHO or surveyor might find useful. These are discussed here briefly so that non-technicians also have some understanding.

Moisture or damp meters (see Figure A1.1) are used to measure the amount of water within a material sample. This measurement allows the user to find out whether the moisture levels are appropriate but more particularly the pattern and location of "dampness" in materials other than timber can be indicative of the type of dampness. Readings are taken on a reference scale as it is only a relative degree of dampness. The wood moisture scale or wood moisture equivalent (WME)is a scale of measuring moisture content in materials as if they were wood and is the most commonly used. The readings indicate whether timber elements are at risk of rot due to dampness (moisture content of less than 17/18% is safe from decay) but timber also needs to contain moisture for its strength. Although damp (excess moisture) encourages the germination of dry rot fungus (*Serpula lacrymans*), that fungus extracts the moisture out of the timber to the extent that it loses its strength.

The first readings will normally be at the surface indicating dampness immediate area of contact of the electrodes. Determining the sub-surface moisture readings will require use of deep wall probes, which entail drilling into the structure and taking readings at increasing depths. The search mode with some models provides a non-invasive technique for scanning walls and floors for moisture without damaging the surface. The search mode

Figure A1.1 Damp meter being used and in this case of rising damp due to bridged damp proof course.

determines the moisture content from the returned signal strength. Condensation will normally only involve damp readings at the surface.

Hygrometers are useful measuring instruments that can be used for ascertaining the RH at the time of inspection. Digital hygrometers are said to be more accurate and some also incorporate a thermometer.

The usefulness of equipment depends on a thorough understanding of what they tell you and the limits of readings.

Additionally, binoculars can be useful when trying to inspect pitched roofs and eaves gutters from the ground.

Dampness Checklist

Fully and properly assessing the cause and identifying the most appropriate solution is a specialist job – a surveyor, a heating engineer, and an environmental health practitioner. But here we give a basic checklist that can help anyone to understand the possible causes and remedies.

The first stage is to carry out a thorough visible check internally to look for the characteristics, and the signs of the problem, using the checklist below. The code given refers to the probable cause of the dampness, and the next stage in the assessment.

For an explanation of the causes, see *Chapter 2* and a brief outline given in the following boxes.

Code	Cause (Type) of dampness
C	Condensation
P	Penetrating
R	Rising
T	Traumatic

1.   Check Internally and Tick the Visible Sign(s)

Signs	Code
Staining to the ceiling of upper floor room(s).	P/T
Staining to external wall(s) around windows or doors.	P
Staining in patch(es) to external wall(s).	P
Staining to the ceiling of lower floor room.	T
Water leaking from joint (or crack) to waste, soil, or water pipe (i.e. below sink or wash hand basin, or to pipe from WC basin).	T
Staining to the base of external ground floor wall(s) to height of about 1 m (3 feet).	R
Rotting skirting boarding to external ground floor wall(s).	R
Wallpaper (or paint) peeling off at base of external ground floor wall(s).	R

(Continued)

(Continued)

Signs	Code
Floor covering to ground floor room(s) lifting.	R
Misting of windows.	C
Wall has "misty" surface.	C
Staining above window or door opening(s).	C
Damp patches with no definite edges.	C
Mould growth to external wall(s) and/or ceiling(s).	C
Water collecting on internal window sills.	C
Staining in patch(es) at the junction of external walls.	C
Stains or streaks on wall particularly below windows (or in bathrooms and kitchens).	
Moisture of surfaces of tiles (or other hard surfaces).	C
Peeling wallpaper generally.	C
Damp or mouldy clothes.	C
Dampness and/or mould behind furniture (such as wardrobes, chests of drawers, and sofas) or equipment (fridges and freezers).	C

In practice, visual inspection cannot diagnose the cause definitively more information will be required.

2. The Next Stage in Trying to Find the Cause Is to Look for the Possible Problems

Code	Probable cause of the damp
C	**Condensation.** This is when warm moisture-laden air is cooled by a relatively cold surface. The surface will feel damp (but this doesn't extend into the fabric of the building).
	The moisture will be at its worst during the colder periods of the year, and mould is most likely for this form of dampness. The "condensation season" is usually October to April in the UK and so the time of year of inspection is an important consideration. Although if mould has not been removed, it will remain outside the "condensation season".
P	**Penetrating.** Rainwater penetrates through the external surface(s) of the dwelling.
	This form of dampness will be worse when there is or has been raining. It may also occur when the snow melts.
R	**Rising.** Where moisture from the soil rises up the wall or floor because of capillary attraction.
	This form of dampness will not be affected by rain and will be fairly constant throughout the year, only being drier during long spells of hot weather. There will often be a "tidemark" at the height of the dampness.
T	**Traumatic.** This is the result of a leak from, or bust to, a water, drainage, or sewage pipe, or a water storage tank.
	If the problem is to a ceiling or internal wall other than those to the top floor, then the dampness may be noticeably worse when a particular facility is used.

3.   Next, Check for Other Signs That Suggest What May Be the Possible Cause

Code	Signs	Comment
C	The dampness will not be affected by rain but may be most noticeable on the inside of the most exposed walls (such as those facing South through to West). It may be (or have started) in areas where there is "cold bridging". The height of the dwelling might be a consideration (See Figure 1.1) as high-rise dwellings might be susceptible to a cooling by the wind and the altitude of the building affects the heat-loss; for every 100 m above sea level, the external temperature reduces by − 0.60°C. If properties are close to the coast, or in an elevated exposed position, it is necessary to add 10% to the heat loss.	This is most likely to occur where the heating system doesn't extend to all parts of the dwelling, where there are problems with the heating system (such as disconnection), or where portable flueless gas or oil heaters are used. Is the heating system adequate for the dwelling and what is the heating pattern (and running costs) of the heating system?
P	The site of the internal damp will give an indication of where the outside problem is likely to be – e.g. if the staining is to a ceiling to the top floor room, then the problem is likely to be the roof; if the staining is to a wall, then it could be damage to the wall such as cracked or missing render, defective rainwater goods. For more details and examples, see subsequently.	
	Slipped, cracked, or missing slate or tile to a pitched roof; or cracked or deteriorated flashing between the roof and chimney stack or vent pipe penetrating the roof (See Figures A1.2 and A1.3). (NB: Usually, it will not be possible to check the state of a flat roof.)	This could allow rainwater (or water from melting snow) to penetrate into the roof space and cause dampness to a ceiling to an upper floor room or landing.
	Leaking rainwater eaves gutter, downpipe, or soil pipe.	There may be staining around the leak (most probably at a joint) and staining to the wall adjacent to or below the leak.
	Cracked or missing external render (coating to the external wall).	This will allow moisture to penetrate into and through the wall.
	Open and perished pointing to a solid wall.	This will allow moisture to penetrate and stain the walls corresponding to the open pointing, and may be associated with a leaking rainwater pipe.

(Continued)

(Continued)

Code	Signs	Comment
	Loose or missing external sealing around a door or window frame and the opening.	This would allow moisture to bypass the frame and cause dampness to the plaster around the reveals of the door or window opening.
	Peeling paintwork and/or rotting timber to window or door frames.	This will allow water to pass through into the interior (if not now, eventually).
R	Staining to the lowest surfaces of external walls.	There may be similar staining to that appearing inside – staining to a height of about 1 m (3 feet), perhaps with a "tide mark" at top.
T	Staining to ceilings and/or walls internally, and may be worse when a particular facility is used but is not affected by the weather.	In this case, the problem will be from an internal source. The pipes to and from facilities should be checked, including wash basins, baths, showers, WC cisterns, and soil pipes. If the staining is to the ceiling to the top floor, this may mean a leak or burst to a tank or pipe in the roof space.

Figures A1.2 and A1.3 The above looked like condensation at first but more detailed investigation revealed defects in the flashings and pointing of the redundant chimney stack above leading to penetrating dampness.

Figures A1.2 and A1.3 (Continued)

4.   What Needs to Be Done?

Here, we give, briefly, the possible remedies.

Code Possible solution(s)

C   Oil or gas flueless heaters should not be used (except in an emergency such as
        a power cut) as these put water vapour (and other gases) into the atmosphere.
        The same applies to gas cookers (hob and oven). These should not be used
        for heating.
        The best option would be to involve a heating engineer to determine and advise
        on the most appropriate solutions. The energy consumption and costs should
        be checked to ascertain how the heating system is used and to give some
        indication as to its efficiency. Sutherland Tables provide comparative costs
        for space heating and hot water for the most common fuels across a range of
        standard house types throughout the UK and Ireland.
        The space heating system should be thoroughly checked for any defects, and to
        see if it can be improved to provide more heat. Is it of adequate size and type
        for the form of construction? If it heats the air up quickly but the structure
        responds only slowly (even if it has good thermal capacity), then the risk of
        condensation is increased.
        Is it capable of heating the whole dwelling economically to a healthy
        temperature and maintaining that temperature?
        It may be necessary to replace the system with one more efficient and one
        providing more heat output to the whole of the dwelling.
        If the windows are single glazed (only one pane of glass), these should be
        replaced with double (or even triple) glazed units. Such units should
        incorporate trickle ventilators.

                                                                    (*Continued*)

(Continued)

Code	Possible solution(s)

Fitting extractor fans or other mechanical ventilation to bathrooms and kitchens can take moisture-laden air outside helping to reduce the total amount of water vapour within the dwelling. To get adequate ventilation without loss of warm air installation of a heat recovery ventilation (HRV) system, sometimes called mechanical ventilation with heat recovery (MVHR) might be appropriate. Rather than just extracting air and replacing it with the air from outside, a heat recovery system draws the heat from the extracted air and passes it to the air which is filtered in from outside[3].

If possible, extra insulation should be added to the external walls. If the walls are of cavity construction (i.e. constructed with two leaves of brick or stone work, with a gap between the leaves), then insulation should be inserted to fill the cavity. Adding insulation to a solid wall is problematic – ideally, it should be added to the outer face (although this may not be possible if the dwelling abuts a footpath or public area); if added to the internal face, then it will reduce the size of the rooms (albeit slightly, by say 50 mm – (2 inches) to the inner surface of external walls).

Any cold bridges should be addressed where possible, but this may not be the case where, for example a concrete balcony slab is a continuation of the concrete floor slab within the dwelling. Sometimes, retro-filling cavities can increase the risk of a cold bridge.

It is sometimes suggested that dehumidifiers are a solution, but they use energy, can be noisy and do not address the underlying problem.

P | Any defects/disrepair to the exterior should be remedied – the structure should be weather-proof and provide protection for those within the dwelling. This includes ensuring that pointing in solid wall construction is sound.

Slipped, cracked, or missing slates or tiles must be replaced and properly fixed. And any flashing sealing the junction between the roof and a chimney stack or vent pipe should be checked, and, if necessary, replaced.

The seal between window and door frames and the surrounding reveals must be sound, and if missing or cracked replaced with an appropriate sealant.

Any rotting woodwork to the window and/or door frames must be replaced, and all timber protected with an appropriate external paint.

Rainwater goods (eaves gutters and down pipes) should be checked for blockages. While it may be possible to carry out a temporary repair to a leak to a gutter, a downpipe or soil pipe, will ultimately, have to be properly fixed or replaced.

Render – is there as a waterproof protective finish, and when it is cracked or a piece is missing, water will get between the render and the wall structure, loosening the render and making it ineffective. This means that the area around the crack or missing part must be removed until a secure render is found, and then the exposed area re-rendered. Alternatively (and probably better) would be to remove and replace the whole of the render to the affected wall.

R | If the damp-stained area of the wall has not deteriorated (e.g. flaking bricks or missing mortar joints), then there will be no need to take any remedial action to the affected wall structure. If there is any deterioration, this should be made good – flaking bricks cut out and renewed, and mortar joints renewed.

(*Continued*)

(Continued)

Code	Possible solution(s)
	To stop any heavy rain "bouncing" up onto the wall, an existing damp-proof course (dpc) must be at least 150 mm (6 inches) above the adjacent ground surface (whether it is soil or a hard surface). This means that the adjacent ground (whatever it is) must be lowered. If necessary, this can be a sort of trench of at least 300 mm (1 foot) wide.
	If there is a render (either from above or from below) extending over a DPC, this must be removed, as it will provide a bridge for moisture to bypass the DPC.
	Where there is no sign of a damp proof course, then a new one must be inserted. This is specialist work and may involve injecting a chemical into the wall to give a waterproof barrier, or the wall cut at a mortar joint and a bitumastic dpc inserted.
T	Any leaking pipe or tank must be removed and properly replaced. This should be done by a specialist (a plumber).

References and Notes

1 A cavity wall is one with a cavity between two leaves, linked together with wall ties. The outer face of a cavity wall will have all bricks lengthwise. A solid wall (9 inches (225mm) thick in older constructions has no cavity

2 This is referred to as a "cold bridge" or "thermal bridge" that occur within the building fabric where, because of the geometry such as corners (junctions between ceiling and wall for example which is why this is where mould might first appear) or the presence of a component of the structure with higher thermal conductivity than the surrounding materials, thereby allowing heat to escape (See Figure 1.1).

3 These systems can be for the whole dwelling or one or two rooms, the main components are a heat recovery cell (in this the indoor heat is extracted and passed on to the external air which is drawn into the building), fans (pulling in external air and blowing out the moisture laden air), and filters to filter out dust and particles from the outside air.

Annex 2
Heat Loss

This book is not intended as detailed technical text, but it is thought that it might be helpful to include more technical information than in the first edition, for those carrying out inspections and those instructing them such as advisers and lawyers.

As has been made clear in Chapter 2, there are four main and inter-related factors that influence the occurrence of condensation

- the moisture generated within the dwelling
- the space heating system provided, and the use made of it
- the ventilation system provided, and the use made of it so that moisture-laden air can be removed from the dwelling
- the thermal capacity, response, and insulation of the dwelling (with adequate insulation, warm moist air cannot come into contact with colder surfaces so cannot be cooled below its dew-point).

It is apparent from that list that some of these factors are the responsibility of the owner or landlord.

When investigating condensation in detail and assessing whether the heating system is adequate and affordable for the property and its occupiers, it might be necessary to look at heat loss. This might entail a heat loss calculation (perhaps also relevant when considering the HHSRS of excess cold). Heat loss is typically measured in kilowatts (kW). This will also require a thorough inspection and identification of the elements of construction.

Where there is a difference in temperatures between two bodies heat flows from the body at higher temperature to that with the lower temperature. The latter gains heat while the former loses heat. This happens to the end when both bodies achieve thermal equilibrium. A building may lose heat through either conduction through the building envelopes (exterior surfaces like walls, floor, roof, windows, and so on) – transmission or fabric heat loss. There is also air infiltration or ventilation losses, whereby cold air from the exterior replaces warmer air inside the property. Thus, draughts have a cooling effect.

A number of factors have to be considered. The surface area of the body is important, and the greater the surface area, the more heat is lost. The speed of heat loss reflects the difference between the bodies, so the greater the difference in temperature, the faster heat is transferred through the structure and more heat input is required to maintain a healthy indoor temperature.

Typically, older forms of construction are more susceptible to heat loss and can be due to a combination of inadequate insulation, the presence of cold (thermal) bridges across the building envelope, single glazing, and lack of air tightness and lack of airtightness (or presence of draughts), for example. The extent of heat loss will vary depending on the type of dwelling so that, for example a mid-terraced house will lose a greater proportion of its heat through the floor and roof than through the walls as it has only two external, which would not be the case for a detached house with the same wall construction, whereas the main route for heat loss from a flat in the middle of a block will be through the external walls.

An important consideration is the term "U-value" – this is the measure of the overall rate of heat transfer through a particular section of construction. U-value is sometimes known as thermal transmittance and indicates the rate at which a building component transmits heat from the inside to the outside. The lower the value, the less heat is transmitted or transferred through the component. U values are measured in watts per square metre per Kelvin [1](W/m^2K) and some typical U values are given in Table A2.1.

Thermal resistance or "r-value" is the measure of the ability of a building material to resist heat flow. It is measured in m^2K/W and in this case the higher the number, the better.

Table A2.1 Typical U values and requirements of current Building Regulations for new construction[1]

Material	U value W/m^2K
Solid brick wall	2
Cavity wall no insulation	1.5
Single glazing	4.8 to 5.8
Double glazing	1.2 to 3.7
Solid Concrete:	3
New roofs	0.16
New walls	0.26
New Floors	0.18
New windows	1.6

[1]Approved Document L: Conservation of Fuel and Power, Vol 1 Dwellings https://assets.publishing.service.gov.uk/media/662a2e3e55e1582b6ca7e592/Approved_Document_L__Conservation_of_fuel_and_power__Volume_1_Dwellings__2021_edition_incorporating_2023_amendments.pdf

Heat loss calculations help determine the energy input necessary to maintain a healthy indoor temperature (and also help design an adequate heating system) and the costs of heating. They also help determine what improvements to insulation can be achieved to reduce heat loss as they highlight where the most effective improvements can be made. When assessing the costs of different forms of heating, Sutherland Tables provide comparative costs for space and water heating[2] and can be useful in assessing whether the dwelling can be maintained economically at healthy temperatures.

Understanding the heat loss and the capacity (rating) of the existing heating system will also allow a better understanding of the adequacy of the heating system to permit healthy temperatures to be maintained economically. As the 2006 HHSRS Operating Guidance says at Box 9[3] on functions and requirements of elements "The dwelling should be provided with adequate thermal insulation and a suitable and effective means of space heating so that the dwelling space can be economically maintained at reasonable temperatures".

Poor workmanship can result in reduced thermal resistance, as can poor junctions or jointing between different materials or what is known as poor detailing and the presence of moisture in insulation.

Calculating Heat Loss

The heat loss from the fabric is the sum of the heat loss through the building elements:

(1) Fabric heat loss = *Area of building component* x *U value* x *temperature difference*
(2) Ventilation heat loss = *Volume of dwelling* x *watts required to raise temperature* x *air change rate* x *temperature difference*.

The heat required to raise 1 cubic metre of air through 1K is taken as 0.33W.

Heat loss is expressed in Watts per K (WK^{-1} or W/K).

Additionally, there may be extra heat loss from thermal bridges. According to the BRE, as dwellings become better insulated, the importance of thermal bridging has increased. In very well-insulated dwellings, the effect that thermal bridging can have on the overall thermal performance of a dwelling can be significant. Research undertaken has shown that thermal bridging can be responsible for up to 30% of a dwelling's heat loss. The heat loss associated with these thermal bridges is expressed as a linear thermal transmittance (Ψ-value).[4] This value is added to the other heat losses to provide the total heat loss.

Over a heating season of say 33 weeks (5,544 hours), the total heat lost would be calculated by multiplying the heat loss by the number of hours of the heating season.

This annex is not intended as comprehensive but sets out the basic principles. A relatively simple example of a heat loss calculation is given here: https://www.open.edu/openlearn/nature-environment/energy-buildings/content-section-2.4.1. The assessment of the adequacy of the heating system can be calculated thus; assuming at the coldest day of the year, the temperature difference between the external air and a healthy indoor temperature is taken as 24°C and total heat loss is calculated as 1500 WK^{-1} the heating system would need to be capable of producing (heat output) 36kW (24 x 1500).

Notes and references

1 The Kelvin, symbol K, is the SI unit of measurement for temperature. Any temperature in degrees Celsius can be converted to kelvin by adding 273.15, The Kelvin scale is designed so one degree Celsius, and one kelvin are the same.

2 https://www.sutherlandtables.co.uk and it is understood these were used in the case *Liverpool City Council v Kassim* [2011] UKUT 169 (LC).

3 Office of the Deputy Prime Minister. (2006) *Housing health and safety rating system operating guidance Housing Act 2004 Guidance about inspections and assessment of hazards given under Section 9*, at page 27, London.

4 See BR 497 is referenced within England and Wales Building Regulations Approved Document Part L, and Scottish Technical Standards Section 6 (Energy).

Further Reading

Battersby SA, Ed (2022). *Clay's handbook of environmental health*, 22nd Edn, Routledge, (Chapter 9).

BS5250:2021. *Management of moisture in buildings – code of practice*, British Standards Institution.

Center for Disease Control and Prevention (2020). *Mold*. Available at https://www.cdc.gov/mold-health/about/index.html?CDC_AAref_Val=https://www.cdc.gov/mold/faqs.htm, and also https://www.cdc.gov/mold-health/communication-resources/

Global Initiative for Asthma (GINA) (2024). *Global Strategy for Asthma Management and Prevention*, see https://ginasthma.org/2024-report/

Housing Ombudsman (2021). *Spotlight on: Damp and mould – It's not lifestyle*. Available at https://www.housing-ombudsman.org.uk/wp-content/uploads/2021/10/Spotlight-report-Damp-and-mould-final.pdf

Institute of Medicine (2004). *Damp indoor spaces and health*. Available at https://www.nap.edu/catalog/11011/damp-indoor-spaces-and-health

London Damp and Mould assessment toolkit. Available at https://www.mecclink.co.uk/london/housing-damp-and-mould-advice/

Marshall D, Worthing D, Dann N, and Health R. (2013). *The construction of houses*, 5th Edn, Routledge, Abingdon, Oxon, UK.

Marshall D, Worthing D, Heath R, and Dann N. (2014). *Understanding housing defects*, 4th Edn, Routledge, Abingdon, Oxon, UK.

National Health Service (2018). *Can damp and mould affect your health*. Available at https://www.nhs.uk/common-health-questions/lifestyle/ can-damp-and-mould-affect-my-health/

NHS Scotland, *Damp and mould indoors*. Available at https://www.nhsinform.scot/healthy-living/indoor-health/damp-and-mould-indoors/

Ormandy D. (2023a). *Condensation and mould: Guidance for housing lawyers*, Legal Action, March 2023, LAG, London.

Ormandy D. (2023b). *Fuel poverty v energy injustice*, Legal Action, April 2023, LAG, London.

Trotman P, Sanders C, and Harrison H. (2004). *Understanding dampness – effects, causes, diagnosis and remedies*, Building Research Establishment. Available at https://www.brebookshop.com/details.jsp?id=148923

UK Health Security Agency, *Spotlight on: The burden of disease caused by damp and mould in English housing*. Available at https://research.ukhsa.gov.uk/our-research/damp-and-mould/

Index